高等应用型人才培养规划教材

Java 程序设计实验指导

（双语版）

主　编　高永平

副主编　何月顺　官芬芬

電子工業出版社

Publishing House of Electronics Industry

北京・BEIJING

内 容 简 介

本书是一部双语实验教材，在教材中提供了相应内容的英文和中文；每一章开始首先对本章主要的知识点给出了概括和精辟的讲解与分析，同时给出 1～2 个综合性的例题；在实验部分围绕本章的重点、难点知识合理选择实验题目，同时给出了实验的参考答案和运行结果。本教材实验内容涵盖广泛，重点培养学生的程序设计能力，列举的例题和实验题目紧密联系实际问题(银行账户问题、银行支票问题、信用卡问题、事务处理、多线程同步以及死锁问题等)，实用性很强。

本教材适合开设该课程的研究生、本科以及高职等各层次人才作为学习 Java 程序设计的实验教材，也可供参加自学考试的学生以及 Java 语言的自学者学习使用。

图书在版编目（CIP）数据

Java 程序设计实验指导：双语版：汉英对照 / 高永平主编. —北京：电子工业出版社，2017.2

ISBN 978-7-121-30676-1

Ⅰ. ①J… Ⅱ. ①高… Ⅲ. ①JAVA 语言—程序设计—高等学校—教学参考资料—汉、英 Ⅳ. ①TP312.8

中国版本图书馆 CIP 数据核字（2016）第 311319 号

策划编辑：程超群　魏建波
责任编辑：郝黎明
印　　刷：北京盛通商印快线网络科技有限公司
装　　订：北京盛通商印快线网络科技有限公司
出版发行：电子工业出版社
　　　　　北京市海淀区万寿路 173 信箱　邮编　100036
开　　本：787×1 092　1/16　印张：10.75　字数：330 千字
版　　次：2017 年 2 月第 1 版
印　　次：2021 年 3 月第 4 次印刷
定　　价：34.50 元

凡所购买电子工业出版社图书有缺损问题，请向购买书店调换。若书店售缺，请与本社发行部联系，联系及邮购电话：（010）88254888，88258888。

质量投诉请发邮件至 zlts@phei.com.cn，盗版侵权举报请发邮件至 dbqq@phei.com.cn。

本书咨询联系方式：（010）88254577，ccq@phei.com.cn。

前　　言

Java 是一门面向对象编程语言，Java 具有简单性、面向对象、分布式、健壮性、安全性、平台独立与可移植性、多线程、动态性等特点。Java 可以编写桌面应用程序、Web 应用程序、分布式系统和嵌入式系统应用程序等。

Java 程序设计是实践性很强的课程，学习 Java 的一个有效方法是多上机实践。本书作者在编写过程中力求概念清晰、通俗易懂，使复杂问题简单化，取材新颖，本书从实际教学出发，加强了对 Java 语言的重点和难点的知识讲解，在实践过程中，充分选用紧密联系实际的问题来深化学生对理论知识的认识，使学生掌握 Java 语言的程序设计的基本方法，让学生基本具备使用 Java 开发实际系统的能力，并培养学生解决实际问题的能力。

本书主要涵盖了 Decisions（选择结构）、Loops（循环结构）、Methods（函数）、Arrays and ArrayLists（数组与数组列表）、Input Output and Exception Handling（输入输出与异常处理）、Objects and Classes（类与对象）、Inheritance and Interfaces（继承与接口）、Graphical User Interfaces（图形用户接口）、the Java Collections Framework（Java 集合框架）、Streams and Binary Input/Output（流与二进制的输入/输出）、Multithreading（多线程）、Programming with JDBC（JDBC 编程）、Internet Networking（Internet 网络）等 13 个章节的内容，每章都从 Key points of this chapter（本章要点）、Example（例题）、Experimental contents（实验内容）、Experimental steps（实验步骤）、Experimental result（实验结果）等四个部分进行介绍。书中所有程序均在 JDK8.0 开发包下编译通过并正确运行；同时在书后面还附加了两套全英文的试卷和参考答案。

为了使学生在上机实验时目标明确以及方便各院校开展双语教学，本实验指导书针对课程内容编写了 13 个实验的中英内容。学生可以在实验课时先仔细查看指导书中给出的本章要点、例题，在此基础上再编写实验程序。为了方便不同背景和实验学时的学生使用，每个实验都是独立性的实验，在教学过程中，教师可以根据实际情况进行适当的裁剪。同时为了方便双语教学的学习和开展，在本书的最后给出了 2 套英文的考试试卷，方便学生对专业知识和英语能力的测试。

在长期的 Java 语言教学过程中，我们发现学生在理论课堂学习的知识总是不能有效地应用于实际编程中，对于遇到的许多问题无从下手，影响了学习效果。本书详细地给出了相应的实验步骤，可引导读者在课后一步一步、循序渐进地完成操作，同时指出了相应的知识要点，难度适中，可以激发读者的学习兴趣，并为以后更深入地学习 Java 程序设计打下扎实的基础。

最后本书在编写的过程中，PRABAHARAN KRISHNAMOORTHY、SIVAPRAKASAM MANIKANDAN、李祥、徐洪珍、张军、王强、汪雪元、吴光明、贾惠珍、章伟、吴建东、姜林、汪宇玲、蔡友林等同行专家为本书提供了大量的素材，并为本书的编写提供了宝贵的意见和建议，丁琪琪、吴深深、欧阳浩、朱芮、于浩瀚、欧阳浩、吴文彬、张特等多位同学参与了本书的资料整理工作，在此一并感谢。

由于作者水平有限，书中难免有疏漏和不妥之处，竭诚希望广大读者和同行专家批评指正。

编者

目　录

第 1 章　Decisions（选择结构）

1.1　Key points of this chapter（本章要点）

1.	The if Statement：The if statement is used to implement a decision. When a condition is fulfilled, one set of statements is executed. Otherwise, another set of statements is executed. if 语句：使用 if 语句来实现一个决定。当条件满足时，执行一组语句。否则，执行另一组语句。

```
if (condition)
{
statements
}
if (condition) { statements1 }
else { statements2 }
```

2.	Nested Branches：It is often necessary to include an if statement inside another. Such an arrangement is called a nested set of statements. 嵌套分支：将一个 if 语句包含在另一个 if 语句内，这通常是必要的。这样的结构被称为嵌套的语句。

1.2　Example（例题）

Here is a typical example: In the United States, different tax rates are used depending on the taxpayer's marital status. There are different tax schedules for single and for married taxpayers. Married taxpayers add their income together and pay taxes on the total. Table 1 gives the tax rate computations. A different tax rate applies to each "bracket". In this schedule, the income in the first bracket is taxed at 10 percent, and the income in the second bracket is taxed at 25 percent. The income limits for each bracket depend on the marital status.

一个典型的例子：在美国，不同的税率是根据纳税人的婚姻状况而决定的。单身和已婚的纳税人有不同的税收计划。已婚的纳税人将他们的收入加在一起，并支付总的税。表 1-1 给出了税率计算。不同税率适用于每一个“分支”。在表中，第一个分支的收入为 10%，第二分支的收入为百分之 25%。每个分支的收入范围取决于婚姻状况。

税率表见表 1-1。

Table1-1 Tax Rate Schedule

If your status is Single and if the taxable income is	the tax is	of the amount over
at most $32,000	10%	$0
over $32,000	$3,200 + 25%	$32,000
If your status is Married and if the taxable income is	the tax is	of the amount over
at most $64,000	10%	$0
over $64,000	$6,400 + 25%	$64,000

```
import java.util.Scanner;
/**
  This program computes income taxes, using a simplified tax schedule.
*/
public class TaxCalculator
{
  public static void main(String[] args)
  {
    final double RATE1 = 0.10;
    final double RATE2 = 0.25;
    final double RATE1_SINGLE_LIMIT = 32000;
    final double RATE1_MARRIED_LIMIT = 64000;
    double tax1 = 0;
    double tax2 = 0;
    // Read income and marital status
    Scanner in = new Scanner(System.in);
    System.out.print("Please enter your income: ");
    double income = in.nextDouble();
    System.out.print("Please enter s for single, m for married: ");
    String maritalStatus = in.next();
    // Compute taxes due
    if (maritalStatus.equals("s"))
    {
      if (income <= RATE1_SINGLE_LIMIT)
      {
        tax1 = RATE1 * income;
      }
      else
      {
        tax1 = RATE1 * RATE1_SINGLE_LIMIT;
        tax2 = RATE2 * (income - RATE1_SINGLE_LIMIT);
      }
    }
    else
    {
      if (income <= RATE1_MARRIED_LIMIT)
      {
        tax1 = RATE1 * income;
```

```
            }
            else
            {
               tax1 = RATE1 * RATE1_MARRIED_LIMIT;
               tax2 = RATE2 * (income - RATE1_MARRIED_LIMIT);
            }
         }
         double totalTax = tax1 + tax2;
         System.out.println("The tax is $" + totalTax);
      }
   }
```

1.3 Experimental contents（实验内容）

<table>
<tr><td rowspan="2">1.</td><td>Write a program that reads a number between 1,000 and 999,999 from the user, where the user enters a comma in the input. Then print the number without a comma. Here is a sample dialog; the user input is in color:
Please enter an integer between 1,000 and 999,999: 23,456
23456
Hint: Read the input as a string. Measure the length of the string. Suppose it contains n characters. Then extract substrings consisting of the first n – 4 characters and the last three characters.</td></tr>
<tr><td>编写一个程序，为用户读取一个介于 1,000 和 999,999 之间的数字，用户在输入数字中输入了一个逗号。然后打印需要没有逗号的数字。下面是一个示例对话框，用户输入的数值为颜色数值：
请输入一个介于 1,000 和 999,999 之间的整数：23,456
23456
提示：读取输入为字符串。计算字符串的长度。假设它包含 n 个字符。然后提取子字符串组成的第 N－4 个字符和最后三个字符。</td></tr>
</table>

<table>
<tr><td rowspan="2">2.</td><td>A supermarket awards coupons depending on how much a customer spends on groceries. For example, if you spend $50, you will get a coupon worth eight percent of that amount. The following table shows the percent used to calculate the coupon awarded for different amounts spent. Write a program that calculates and prints the value of the coupon a person can receive based on groceries purchased.
Here is a sample run:
Please enter the cost of your groceries: 14
You win a discount coupon of $ 1.12. (8% of your purchase)</td></tr>
<tr><td>一个超市奖励券取决于顾客在商品上花费的多少。例如，如果你花 50 美元，你会得到一张价值 8%的优惠券。表 1-2 的折扣表显示用于计算所花费的不同金额的优惠</td></tr>
</table>

券的百分比。编写一个程序，在一个人购买商品的基础上计算和打印优惠券的值。

示例运行结果：

请输入你的商品成本：14

你赢了 1.12 美元的折扣。（你购买的 8%）

Table 1-2　Discount Schedule

Money Spent（花费）	Coupon Percentage（折扣率）
Less than $10	No coupon
From $10 to $60	8%
More than $60 to $150	10%
More than $150 to $210	12%
More than $210	14%

1.4　Experimental steps（实验步骤）

1.

```
import java.util.Scanner;
public class CharactersTranverse {
    public static void main(String[] args) {
        // TODO Auto-generated method stub
        System.out.print("Please enter an integer between 1,000 and
            999,999:");
        Scanner in=new Scanner(System.in);
        String num=in.next();
        in.close();
        String result="";
        while(num.length()>3){
            result=num.substring(num.length()-3)+result;
            num=num.substring(0,num.length()-4);
        }
        result=num+result;
        System.out.print("The result is "+result);
    }
}
```

2.

```
import java.util.Scanner;
public class Coupon {
    public static void main(String[] args) {
        // TODO Auto-generated method stub
        System.out.print("Please enter the cost of your groceries:");
        Scanner in= new Scanner(System.in);
        double cost=in.nextDouble();
        if(cost<=10){
            System.out.print("Sorry,there is no coupon!");
```

```
            }
            else if(cost>10&&cost<=60){
                System.out.print("You win a discount coupon of $"+ cost*0.08+
                    "(8% of your purchase)!");
            }
            else if(cost>60&&cost<=150){
                System.out.print("You win a discount coupon of $"+ cost*0.1+
                    "(10% of your purchase)!");
            }
            else if(cost>150&&cost<=210){
                System.out.print("You win a discount coupon of $"+ cost*0.12+
                    "(12% of your purchase)!");
            }
            else{
                System.out.print("You win a discount coupon of $"+ cost*0.14+
                    "(14% of your purchase)!");
            }
        }
    }
```

1.5 Experimental result（实验结果）

实验结果如图 1-1 和图 1-2 所示。

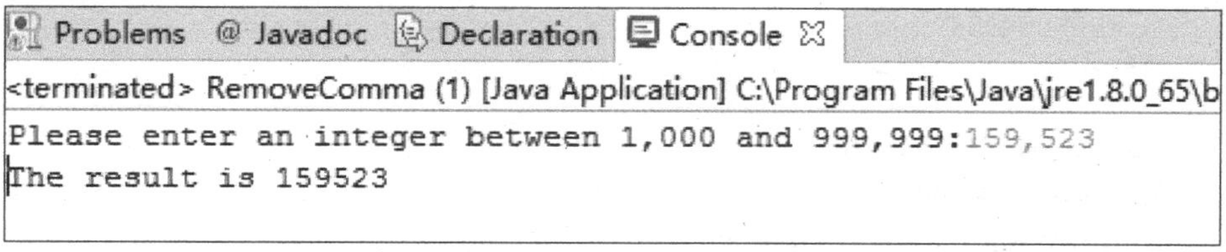

```
Problems  @ Javadoc  Declaration  Console
<terminated> RemoveComma (1) [Java Application] C:\Program Files\Java\jre1.8.0_65\b
Please enter an integer between 1,000 and 999,999:159,523
The result is 159523
```

Fig1-1.　CharactersTranverse

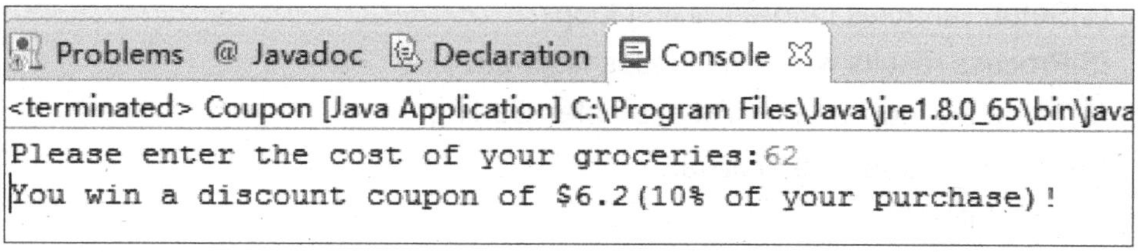

```
Problems  @ Javadoc  Declaration  Console
<terminated> Coupon [Java Application] C:\Program Files\Java\jre1.8.0_65\bin\java
Please enter the cost of your groceries:62
You win a discount coupon of $6.2(10% of your purchase)!
```

Fig1-2.　Coupon

第2章 Loops（循环结构）

2.1 Key points of this chapter（本章要点）

<table>
<tr><td rowspan="2">1.</td><td>The while Loop：In Java, the while statement implements such a repetition. It has the form.</td></tr>
<tr><td>while 循环：在 Java 里面，while 语句实现这样一个重复。它的形式如下：</td></tr>
</table>

```
while (condition)
  {
    statements
  }
```

<table>
<tr><td rowspan="2">2.</td><td>The for Loop：there is a special form for it, called the for loop.</td></tr>
<tr><td>for 循环：这里有一个特殊的形式叫 For 循环。</td></tr>
</table>

```
for (initialization; condition; update)
  {
    statements
  }
```

<table>
<tr><td rowspan="2">3.</td><td>Steps to Writing a Loop
(1)Decide what work to do inside the loop；
(2)Specify the loop condition；
(3)Determine loop type；
(4)Setup variables before the first loop；
(5)Process results when the loop is finished；
(6)Trace the loop with typical examples Coding；
(7)Implement the loop in Java.</td></tr>
<tr><td>循环的步骤
（1）决定在循环里面要做什么；
（2）明确循环条件；
（3）决定采用哪种循环方式；
（4）在第一个循环开始之前设置变量；
（5）在循环结束后对结果进行加工；
（6）用典型的例子代码来跟踪这个循环；
（7）在 Java 里面实现这个循环。</td></tr>
</table>

2.2 Example（例题）

You put $10,000 into a bank account that earns 5 percent interest per year. How many years does it take for the account balance to be double the original investment?

你把 10,000 美元放在银行里，每年的利息是 5%，多少年之后，你的余额会变成最初的 2 倍？

```
/**
  This program computes the time required to double an investment.
这个程序计算要花多少年投资会变成原来的 2 倍。
*/
public class DoubleInvestment
{
   public static void main(String[] args)
   {
      final double RATE = 5;
      final double INITIAL_BALANCE = 10000;
      final double TARGET = 2 * INITIAL_BALANCE;
      double balance = INITIAL_BALANCE;
      int year = 0;
      // Count the years required for the investment to double
      while (balance < TARGET)
      {
         year++;
         double interest = balance * RATE / 100;
         balance = balance + interest;
      }
       System.out.println("The investment doubled after "
         + year + " years.");
   }
}
```

2.3 Experimental contents（实验内容）

Write programs that read a sequence of integer inputs and print:

a. The smallest and largest of the inputs;

b. The number of even and odd inputs;

c. Cumulative totals. For example, if the input is 1 7 2 9, the program should print 1 8 10 19;

d. All adjacent duplicates. For example, if the input is 1 3 3 4 5 5 6 6 6 2, the program should print 3 5 6.

编写一个程序，读输入的一串整型值并按要求输出：

输入的最大值和最小值；

奇数和偶数的个数；

依次累计输入总值的和。例如，如果输入 1 7 2 9，那么程序应该输出 1 8 10 19

输出输入数值串中相邻重复的值。例如，如果输入是 1 3 3 4 5 5 6 6 6 2，那么程序应该输出 3 5 6

2.4 Experimental steps（实验步骤）

1.

```
import java.util.Scanner;
/**
 * The smallest and largest of the inputs:找出输入值中的最大值和最小值
 * @author *****
 *
 */
public class MaxAndMin {
    public static void main(String[] args) {
        // TODO Auto-generated method stub
        System.out.print("请输入一些数值(输入非数值值结束): ");
        Scanner in=new Scanner(System.in);
        boolean run=false;      //监测是否输入值
        int temp;
        int min=Integer.MAX_VALUE;      //使用极大值初始化最小值
        int max=Integer.MIN_VALUE;      //使用极小值初始化最大值
        while(in.hasNextInt()){
            run=true;
            temp=in.nextInt();
            if(temp>max){
                max=temp;
            }
            if(temp<min){
                min=temp;
            }
        }
        in.close();
        if(run){
            System.out.println("最大值为: "+max+",最小值为: "+min);
        }
        else{
            System.out.println("没有输入任何值! ");
        }
    }
}
```

2.

```
import java.util.Scanner;
/**
```

```
 * The number of even and odd inputs:合计输入整数中奇偶数的个数
 * @author *****
 *
 */
public class EvenAndOdd {
    public static void main(String[] args) {
        // TODO Auto-generated method stub
        System.out.print("请输入一些整数(输入非数值值结束): ");
        Scanner in=new Scanner(System.in);
        int temp;
        int NumOfEven=0;               //偶数记数
        int NumOfOdd=0;                //奇数记数
        while(in.hasNextInt()){
            temp=in.nextInt();
            if(temp%2==0){
                NumOfEven++;
            }
            else{
                NumOfOdd++;
            }
        }
        in.close();
        System.out.println("偶数个数为"+NumOfEven+",奇数个数为"+NumOfOdd);
    }
}
```

3.

```
import java.util.Scanner;
/**
 * Cumulative totals:依次累计输入数值的总和
 * @author *****
 *
 */
public class Total {
    public static void main(String[] args) {
        // TODO Auto-generated method stub
        System.out.print("请输入一些整数(输入非数值值结束): ");
        final int MAXLENGTH=20;
        Scanner in=new Scanner(System.in);
        int temp[]=new int[MAXLENGTH];
        int length=0;
        while(in.hasNextInt()){
            temp[length]=in.nextInt();
            length++;
        }
        in.close();
        int total=0;
        for(int i=0;i<length;i++){
            total+=temp[i];
```

```
                System.out.print(total+" ");
            }
        }
    }
```

4.

```
    import java.util.Scanner;
    /**
     * All adjacent duplicates:输出输入数值串中相邻重复的值
     * @author*****
     *
     */
    public class AdjacentDuplicates {
        public static void main(String[] args) {
            // TODO Auto-generated method stub
            System.out.print("请输入一些整数(输入非数值值结束): ");
            final int MAXLENGTH=20;
            Scanner in=new Scanner(System.in);
            int temp[]=new int[MAXLENGTH];
            int length=0;
            while(in.hasNextInt()){
                temp[length]=in.nextInt();
                length++;
            }
            in.close();
            int sentry=temp[0];            //记录上一个值
            boolean isDispaly=false;
            for(int i=1;i<length;i++){
                if(temp[i]==sentry){
                    if(!isDispaly){
                        System.out.print(temp[i]+" ");
                        isDispaly=true;
                    }
                }
                else{
                    sentry=temp[i];
                    isDispaly=false;
                }
            }
        }
    }
```

2.5 Experimental result（实验结果）

实验结果如图 2-1～图 2-4 所示。

```
请输入一些数值(输入非数值值结束): 8 6 2 4 5 -1.
最大值为: 8,最小值为: -1
```

Fig2-1. The smallest and largest of the inputs

```
请输入一些整数(输入非数值值结束)：5 2 6 7 1 .
偶数个数为2,奇数个数为3
```

Fig2-2.　The number of even and odd inputs

```
请输入一些整数(输入非数值值结束)：1 3 5 7 9 .
1 4 9 16 25
```

Fig2-3.　Cumulative totals.

```
请输入一些整数(输入非数值值结束)：1 1 2 3 3 3 5 5 .
1 3 5
```

Fig2-4.　All adjacent duplicates.

第 3 章 Methods（函数）

3.1 Key points of this chapter（本章要点）

1.	A method is a sequence of instructions with a name. 一个方法是一个具有名称的指令序列。
(1)	You declare a method by defining a named block of code. 通过定义一个已命名的代码块来声明一个方法。

```
public static void main(String[] args)
{
  double result = Math.pow(2, 3);
  . . .
}
```

(2)	You call a method in order to execute its instructions. 你调用一个方法来执行其指令。

(3)	Method Declaration 方法声明

方法声明如图 3-1 所示。

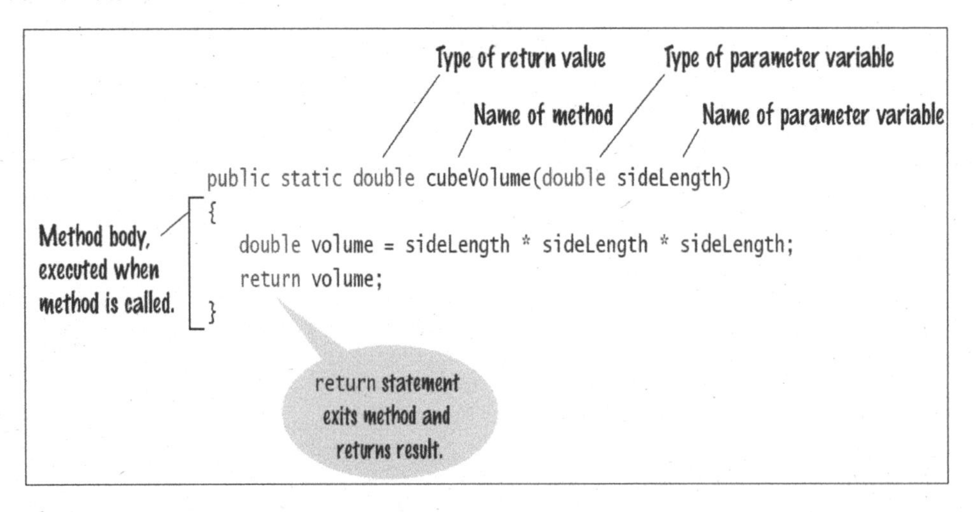

Fig3-1. Method Declaration

2. Parameter Passing（参数传递）

	Parameter variables receive the argument values supplied in the method call. 参数变量得到方法调用中提供的参数值。

参数传递如图 3-2 所示。

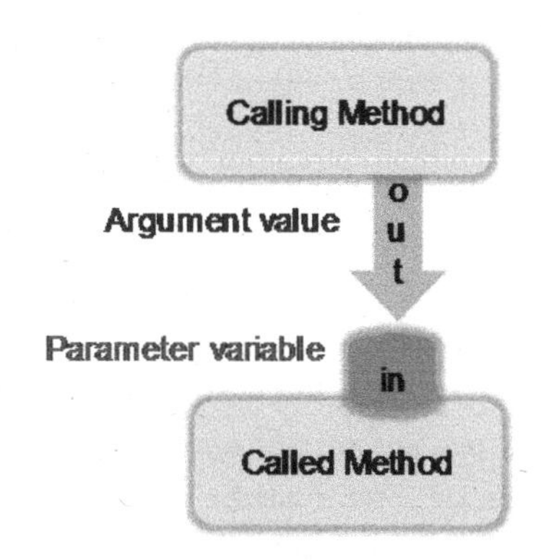

Fig3-2.　Parameter Passing

3. Return Values（返回值）

	Methods can (optionally) return one value：Declare a return type in the method declaration；Add a return statement that returns a value。
	方法可以（可选的）返回一个值：声明一个方法声明的返回类型;添加一个返回语句使得返回一个值。

3.2　Example（例题）

Let us apply the process of stepwise refinement to a programming problem. When printing a check, it is customary to write the check amount both as a number (“$274.15”) and as a text string (“two hundred seventy four dollars and 15 cents”). Doing so reduces the recipient's temptation to add a few digits in front of the amount.
让我们将逐步求精的过程应用到一个编程问题中。打印支票时，习惯上写支票金额作为数字（“$274.15”）和作为一个文本字符串（“二百七十四美元 15 美分”）。这样做可以打消接收者情不自禁地在数额前面添加几个数字。
For a human, this isn't particularly difficult, but how can a computer do this? There is no built-in method that turns 274 into "two hundred seventy four". We need to program this method. Here is the description of the method we want to write:
对于一个人而言，这并不是格外困难的，但计算机如何能做到这点？没有内置的方法把 274 变成“二百七十四”。我们需要编程设计这个方法，这里是我们想要去编写的方法的描述：

```
import java.util.Scanner;
/**
   This program turns an integer into its English name.
*/
public class IntegerName
{
```

```
    public static void main(String[] args)
    {
        Scanner in = new Scanner(System.in);
        System.out.print("Please enter a positive integer < 1000: ");
        int input = in.nextInt();
        System.out.println(intName(input));
    }
    /**
        Turns a number into its English name.
        @param number a positive integer < 1,000
        @return the name of the number (e.g. "two hundred seventy four")
    */
    public static String intName(int number)
    {
        int part = number; // The part that still needs to be converted
        String name = ""; // The name of the number
        if (part >= 100)
        {
            name = digitName(part / 100) + " hundred";
            part = part % 100;
        }
        if (part >= 20)
        {
            name = name + " " + tensName(part);
            part = part % 10;
        }
        else if (part >= 10)
        {
            name = name + " " + teenName(part);
            part = 0;
        }
        if (part > 0)
        {
            name = name + " " + digitName(part);
        }
        return name;
    }
    /**
        Turns a digit into its English name.
        @param digit an integer between 1 and 9
        @return the name of digit ("one" ... "nine")
    */
    public static String digitName(int digit)
    {
        if (digit == 1) { return "one"; }
        if (digit == 2) { return "two"; }
```

```
        if (digit == 3) { return "three"; }
        if (digit == 4) { return "four"; }
        if (digit == 5) { return "five"; }
        if (digit == 6) { return "six"; }
        if (digit == 7) { return "seven"; }
        if (digit == 8) { return "eight"; }
        if (digit == 9) { return "nine"; }
        return "";
    }
    /**
        Turns a number between 10 and 19 into its English name.
        @param number an integer between 10 and 19
        @return the name of the given number ("ten" ... "nineteen")
    */
    public static String teenName(int number)
    {
        if (number == 10) { return "ten"; }
        if (number == 11) { return "eleven"; }
        if (number == 12) { return "twelve"; }
        if (number == 13) { return "thirteen"; }
        if (number == 14) { return "fourteen"; }
        if (number == 15) { return "fifteen"; }
        if (number == 16) { return "sixteen"; }
        if (number == 17) { return "seventeen"; }
        if (number == 18) { return "eighteen"; }
        if (number == 19) { return "nineteen"; }
        return "";
    }
    /**
        Gives the name of the tens part of a number between 20 and 99.
        @param number an integer between 20 and 99
        @return the name of the tens part of the number ("twenty" ... "ninety")
    */
    public static String tensName(int number)
    {
        if (number >= 90) { return "ninety"; }
        if (number >= 80) { return "eighty"; }
        if (number >= 70) { return "seventy"; }
        if (number >= 60) { return "sixty"; }
        if (number >= 50) { return "fifty"; }
        if (number >= 40) { return "forty"; }
        if (number >= 30) { return "thirty"; }
        if (number >= 20) { return "twenty"; }
        return "";
    }
}
```

3.3 Experimental contents（实验内容）

<table>
<tr><td rowspan="2">1.</td><td>Write a method called printPyramid that is passed an odd integer n and a String s, and that prints a pyramidal shape using s. The top of the pyramid has a single copy of s, and each successive row has two additional copies of s. The last row contains n copies of s. For example, calling printPyramid(21, "*"); prints the following lines:
Test your work by calling printPyramid(21, "*") from the main method.</td></tr>
<tr><td>编写一个 printPyramid 方法用来传递一个奇数 n 和一个字符串 s，并且通过 s 打印出一个金字塔形状。金字塔的顶端有一个单个的 s，每一行增加两个 s。最后一行包含 n 个 s。例如，调用 printPyramid(21, "*")；打印出下面的图形
测试你的工作通过从 main 方法中调用 printPyramid(21, "*")</td></tr>
</table>

<table>
<tr><td rowspan="2">2.</td><td>Now that we can decode single digits, it's time to build some code that will help detect errors in credit card numbers. Here's the idea:</td></tr>
<tr><td>现在我们可以解码一些数字，是时候写一些代码来帮助我们发现信用卡号上的错误，编程思想如下：</td></tr>
<tr><td rowspan="2">a)</td><td>Starting with the check digit and moving left, compute the sum of all the decoded digits.</td></tr>
<tr><td>首先确认数字然后左移，计算所有被解码的数字之和。</td></tr>
<tr><td rowspan="2">b)</td><td>Compute the remainder of the sum using integer division by 10. If the result is not zero, the credit card number is invalid. Otherwise, the card number is likely to be valid.</td></tr>
<tr><td>对刚刚那个结果进行对 10 求余，如果最后的结果不是 0，那么这个信用卡号是无效的，否则，这个信用卡有效。</td></tr>
<tr><td rowspan="2"></td><td>Here are two examples:</td></tr>
<tr><td>这里是两个例子：</td></tr>
</table>

```
Card number: 2315778          Card number 1234567
```

```
        decode(8, false) = 8        decode(7, false) = 7
        decode(7, true)  = 5        decode(6, true)  = 3
        decode(7, false) = 7        decode(5, false) = 5
        decode(5, true)  = 1        decode(4, true)  = 8
        decode(1, false) = 1        decode(3, false) = 3
        decode(3, true)  = 6        decode(2, true)  = 4
        decode(2, false) = 2        decode(1, false) = 1
                Sum = 30                    Sum = 31
           30 mod 10 = 0               31 mod 10 = 1
    This number may be valid    This number is invalid
```

Write a static method called checkDigits that is passed a seven-digit credit card number and that performs the steps described above. Reuse the decode method that you wrote in Lab. The method should return the word “valid” if the number passes the test and “invalid” otherwise.

写一个叫做 checkDigits 的静态方法来检验 7 位数信用卡号是否正确，并用上述的方法进行检测。再使用你在实验上的解码方法。如果计算正确的话，这个方法最后应该返回“有效”，否则返回“无效”。

In order to verify that the card number is correct we will need to decode every digit. The decoding process depends on the position of the digit within the credit card number:

为了验证卡号是正确的，我们需要将每一位数字“解码”。解码过程取决于信用卡号中的数字的位置。

a)	If the digit is in an odd-numbered position, simply return the digit, 如果数字在奇数的位置，则返回数字。
b)	If the digit is in an even-numbered position, double it. If the result is a single digit, return it; otherwise, add the two digits in the number and return the sum. 如果数字在偶数的位置，数字加倍。如果结果是一个数字，返回它；否则，对这两个数字做加法并返回求和的结果。

For example, if we decode 8 and it is in an odd position, we return 8. On the other hand, if 8 is in an even position, we double it to get 16, and then return 1 + 6 = 7. Decoding 4 in an odd position would return 4, and decoding it an even position would return 8.

例如，如果解码 8 并且它在奇数的位置，则返回 8。另一方面，如果 8 是在一个偶数位置，加倍之后得到 16，然后返回 1 + 6 = 7。解码在一个奇数的位置 4 将返回 4，而解码在偶数位置将返回 8。

Test your methods with the main method below:

测试你编写的函数，使用下面的 main 方法。

```
    public class Luhn
```

```
{
  public static void main(String[] args)
  {
    int num = 2315778;
    System.out.println("Credit card number: " + num + " is " +
      checkDigits(num));
    num = 1234567;
    System.out.println("Credit card number: " + num + " is " +
      checkDigits(num));
    num = 7654321;
    System.out.println("Credit card number: " + num + " is " +
      checkDigits(num));
    num = 1111111;
    System.out.println("Credit card number: " + num + " is " +
      checkDigits(num));
  }
  // Put your code here
public static int decode(int digit, boolean even)
  {
    //…………
}
public static String checkDigits(int cardNo)
{
//…………
}
}
```

3.4 Experimental steps（实验步骤）

1.

```
import java.util.Scanner;
/**
 * 按照输入的要求输出金字塔形
 * @author ********
*/
public class printPyramid {
   public static void main(String [] args){
      Scanner in=new Scanner(System.in);
      System.out.print("请输入金字塔基本字符：");
      String s=in.next();
      System.out.print("请输入金字塔底部长度，限奇数：");
      int n=in.nextInt();
      boolean input=false;
```

```
        if(n%2==1){
            input=true;
        }
        while(!input){
            System.out.print("输入数据错误，请输入金字塔底部长度，限奇数：");
            n=in.nextInt();
            if(n%2==1){
                input=true;
            }
        }
        in.close();
        Pyramid(n,s);
    }
    public static void Pyramid (int n,String s){
        for(int i=1;i<(n+3)/2;i++)
        {
            for(int b=0;b<(n+1)/2-i;b++)
            {
                System.out.print(" ");
            }
            for(int j=2;j<2*i+1;j++)
            {
                System.out.print(s);
            }
            System.out.println();
        }
    }
}
```

2.

```
/**
   @author…
*/
public class CreditCardCheck
{
    public static void main(String[] args)
      {
         int num = 2315778;
         System.out.println("Credit card number: " + num + " is " +
           checkDigits(num));
         num = 1234567;
         System.out.println("Credit card number: " + num + " is " +
           checkDigits(num));
         num = 7654321;
         System.out.println("Credit card number: " + num + " is " +
```

```
      checkDigits(num));
    num = 1111111;
    System.out.println("Credit card number: " + num + " is " +
      checkDigits(num));
  }
  public static int decode(int digit, boolean even)
  {
    int temp = 0;
    int digitDoubled = 2 * digit;
    if (even)
    {
      if (digitDoubled > 9)
      {
        temp = (digitDoubled % 10) + (digitDoubled / 10);
      }
      else
      {
        temp = digitDoubled;
      }
    }
    else
    {
      temp = digit;
    }
    return temp;
  }
  public static String checkDigits(int cardNo)
  {
    int sum = 0;
    boolean even = false; // First number is in position 7
    while (cardNo != 0)
    {
      int nextDigit = cardNo % 10;
      cardNo = cardNo / 10;
      sum = sum + decode(nextDigit, even);
      even = !even;
    }
    if (sum % 10 == 0)
    {
      return "valid";
    }
    else
    {
      return "invalid";
    }
```

```
        }
    }
```

3.5　Experimental result（实验结果）

实验结果如图 3-3 和图 3-4 所示。

Fig3-3.　printPyramid

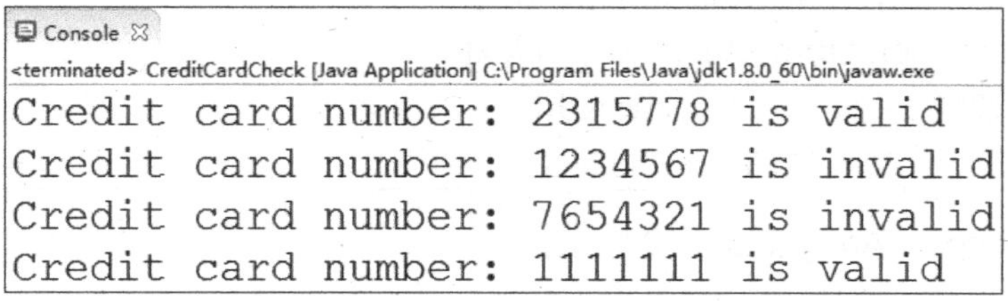

```
Console ⊠
<terminated> CreditCardCheck [Java Application] C:\Program Files\Java\jdk1.8.0_60\bin\javaw.exe
Credit card number: 2315778 is valid
Credit card number: 1234567 is invalid
Credit card number: 7654321 is invalid
Credit card number: 1111111 is valid
```

Fig3-4.　CreditCardCheck

第 4 章 Arrays and ArrayLists（数组与数组列表）

4.1 Key points of this chapter（本章要点）

You need to collect large numbers of values. In Java, you use the array and array list constructs for this purpose. Arrays have a more concise syntax, whereas array lists can automatically grow to any desired size.

在 Java 中，为了保存大量的数值，您可以使用数组和数组列表结构。数组有更简洁的语法，而数组列表可以自动增加到任何想要的大小。

1. Arrays

To construct an array: new typeName[length]

To access an element: arrayReference[index]

构建一个数组：new typeName[length]

访问数组元素：arrayReference[index]

2. Declaring an Array(定义一个数组)

```
double[] values;  values = new  double[10];
double[] values = new double[10];
```

When you declare an array, you can specify the initial values. For example：

当你定义一个数组的时候，你可以指定数组的初始值。例如：

```
double[] moreValues = { 32, 54, 67.5, 29, 35, 80, 115, 44.5, 100, 65 };
```

定义一个数组如图 4-1 所示。

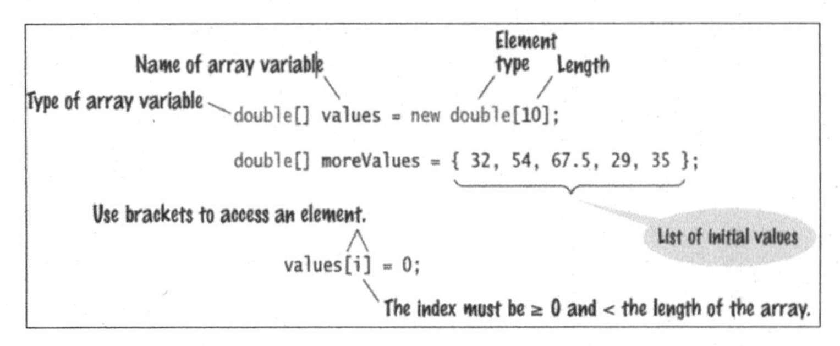

Fig4-1. Declaring an Array

3. The Enhanced for Loop（增强的 for 循环）

Using for loops to 'walk' arrays is very common，For example：

使用 for 循环所有数组是非常普遍的，例如：

```
double[] values = . . .;
    double total = 0;
    for (double element : values)
    {
    total = total + element;
}
```

This loop is equivalent to the following for loop and an explicit index variable:

这个循环相当于下面带有 index 索引变量的循环：

```
for (int i = 0; i < values.length; i++)
{
    double element = values[i];
    total = total + element;
}
```

4. Using Arrays with Methods（在方法中使用数组）

When you define a method with an array argument, you provide a parameter variable for the array. For example, the following method computes the sum of an array of floating-point numbers:

当你在自定的方法中使用数组的内容时，需要提供一个数组变量作为该方法的参数。例如：下面的方法是计算浮点数的数组之和：

```
public static double sum(double[] values)
{
     double total = 0;
     for (double element : values)
     {
     total = total + element;
     }
     return total;
}
```

5. Two-Dimensional Arrays（二维数组）

It often happens that you want to store collections of values that have a twodimensional layout. Such data sets commonly occur in financial and scientific applications. An arrangement consisting of rows and columns of values is called a two-dimensional array, or a matrix.

经常会发生你想保存的数据会呈现出一个二维的布局，这样的数据集一般出现在金融和科学应用中。这种由行、列组成的值称为二维数组或矩阵。

二维数组如图 4-2 所示。

Name　Element type　Number of rows　Numberof columns

```
double[][] tableEntries = new double[7][3];
```

All values are initialized with 0.

Name

List of initial values

```
int[][] data = {
                { 16, 3, 2, 13 },
                { 5, 10, 11, 8 },
                { 9, 6, 7, 12 },
                { 4, 15, 14, 1 },
              };
```

Fig4-2.　Two-Dimensional Array Declaration

6. Array Lists（数组列表）

When you write a program that collects inputs, you don't always know how many inputs you will have. In such a situation, an array list offers two significant advantages: Array lists can grow and shrink as needed;The ArrayList class supplies methods for common tasks, such as inserting and removing elements.

当你编写一个程序保存所输入的内容时，并不是总是能够知道你要输入多少。在这种情况下，数组列表提供了两个重要的优势：数组列表可以根据需要增长和收缩；ArrayList 类为常见的任务提供了方法，比如插入和删除元素。

To construct an array list: new ArrayList<typeName>()
To access an element: arraylistReference.get(index)
arraylistReference.set(index, value)

创建一个数组列表：new ArrayList<typeName>()
访问数组列表的成员：arraylistReference.get(index)
arraylistReference.set(index, value)

数组列表如图 4-3 所示。

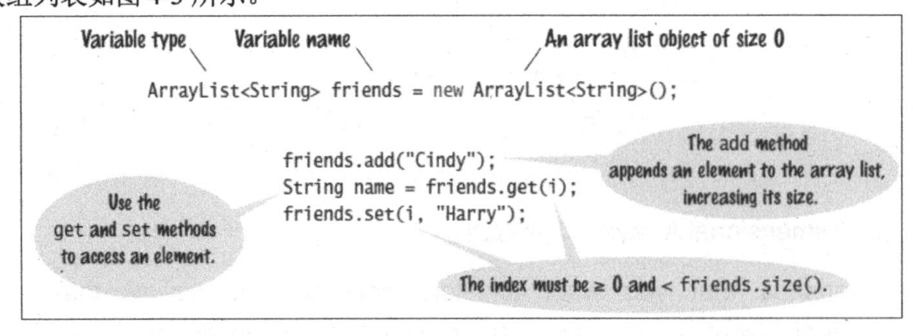

Fig4-3.　Array Lists

4.2　Example（例题）

The following program puts an algorithms to work: to mark the largest value in an input sequence.

下面的程序展示了一个算法：标记输入序列的最大值。

```
import java.util.Scanner;
/**
This program reads a sequence of values and prints them, marking the largest
  value.
这个程序读取并打印一个序列的值,标记出其中的最大值。
*/
public class LargestInArray
{
  public static void main(String[] args)
  {
    final int LENGTH = 100;
    double[] values = new double[LENGTH];
    int currentSize = 0;
    // Read inputs
      //读取输入的数据
    System.out.println("Please enter values, Q to quit:");
    Scanner in = new Scanner(System.in);
    while (in.hasNextDouble() && currentSize < values.length)
    {
      values[currentSize] = in.nextDouble();
      currentSize++;
    }
    // Find the largest value
      //寻找最大值
    double largest = values[0];
    for (int i = 1; i < currentSize; i++)
    {
      if (values[i] > largest)
      {
        largest = values[i];
      }
    }
    // Print all values, marking the largest
      //打印出所有的数值，并且标记出最大值
    for (int i = 0; i < currentSize; i++)
    {
      System.out.print(values[i]);
      if (values[i] == largest)
      {
        System.out.print(" <== largest value");
      }
      System.out.println();
    }
  }
}
```

Here is the same algorithm, now using an array list:

下面的代码使用数组列表实现相同的功能：

```
import java.util.ArrayList;
import java.util.Scanner;
/**
   This program reads a sequence of values and prints them, marking the
     largest value.
*/
/**
   这个程序读取并打印一个序列的值,标记出其中的最大值。*/
public class LargestInArrayList
{
   public static void main(String[] args)
   {
      ArrayList<Double> values = new ArrayList<Double>();
      // Read inputs
         // 读取数据
      System.out.println("Please enter values, Q to quit:");
      Scanner in = new Scanner(System.in);
      while (in.hasNextDouble())
      {
         values.add(in.nextDouble());
      }
      // Find the largest value
        //寻找最大值
      double largest = values.get(0);
      for (int i = 1; i < values.size(); i++)
      {
         if (values.get(i) > largest)
         {
            largest = values.get(i);
         }
      }
      // Print all values, marking the largest
        //打印出所有的数值，标记出最大值
      for (double element : values)
      {
         System.out.print(element);
         if (element == largest)
         {
            System.out.print(" <== largest value");
         }
         System.out.println();
      }
   }
}
```

4.3 Experimental contents（实验内容）

1\. Write a method that is passed an array, x, of doubles and an integer rotation amount, n. The method creates a new array with the items of x moved forward by n positions. Elements that are rotated off the array will appear at the end. For example, suppose x contains the following items in sequence:

1 2 3 4 5 6 7

After rotating by 3, the elements in the new array will appear in this sequence:

4 5 6 7 1 2 3

写一个方法，传入一个 double 类型数组 x 和整形的循环总数 n。通过把 x 向前移动 n 个位置，该方法创建出一个新数组。新数组的元素则在数组最末端循环。例如，x 数组包含以下元素：

1 2 3 4 5 6 7

循环 3 次后，新数组中的元素将是这样的:

4 5 6 7 1 2 3

2\. Array lists are objects that, like arrays, provide you the ability to store items sequentially and recall items by index. Working with array lists involves invoking ArrayList methods, so we will need to develop some basic skills. Let's start with the code below:

数组列表就像数组一样，通过索引按顺序存储数据和回调数据。使用数组列表需要调用 ArrayList 的方法，所以我们需要开发一些基本技能。让我们从下面的代码开始：

```
import java.util.ArrayList;
public class ArrayListRunner
{
   public static void main(String[] args)
   {
      ArrayList<String> names = new ArrayList<String>();
      System.out.println(names);
   }
}
```

The main method imports java.util.ArrayList and creates an ArrayList that can hold strings. It also prints out the ArrayList and, when it does, we see that the list is empty: [ ].

main 函数使用了 java.util.ArrayList 包下的方法并且创建了一个能够保存字符串的数组列表。它还输出 ArrayList，当输出时，我们看到列表为空:[]。

Complete the following tasks by adding code to this skeleton program. If you are asked to print a value, provide a suitable label to identify it when it is printed.

a) Invoke add() to enter the following names in sequence: Alice, Bob, Connie, David,

Edward, Fran, Gomez, Harry. Print the ArrayList again.

b) Use get() to retrieve and print the first and last names.

c) Print the size() of the ArrayList.

d) Use size() to help you print the last name in the list.

e) Use set() to change “Alice” to “Alice B. Toklas”. Print the ArrayList to verify the change.

f) Use the alternate form of add() to insert “Doug” after “David”. Print the ArrayList again.

g) Use an enhanced for loop to print each name in the ArrayList.

h) Create a second ArrayList called names2 that is built by calling the ArrayList constructor that accepts another ArrayList as an argument. Pass names to the constructor to build names2. Then print the ArrayList.

i) Call names.remove(0) to remove the first element. Print names and names2. Verify that Alice B. Toklas was removed from names, but not from names2.

j) Create a third ArrayList called names3, use Clone method to realize cloning names to names3.

通过向这个程序添加代码完成以下任务。如果要求你输入一个值，当它被输出时提供一个合适的标签来识别它。

a）调用 add()依次输入以下名称：Alice，Bob，Connie，David，Edward，Fran，Gomez，Harry。再输出 ArrayList。

b）使用 get()来检索和输出第一个和最后一个名称。

c）输出 ArrayList 的大小。

d）使用 size()输出列表中的姓名。

e）使用 set()把“Alice”更改为“Alice B. Toklas”。输出 ArrayList 中的内容来验证变化。

f）使用 add()把“Doug”插入“David”之后。再次输出 ArrayList 中的内容。

g）使用一个增强的 for 循环输出 ArrayList 中的每个名字。

h）由构造函数调用 ArrayList，接受另一个 ArrayList 作为形参来创建第二个 ArrayList 对象叫做 names2。将 names 作为参数传递给构造函数来构建 names2。然后输出 ArrayList 的内容。

i）调用 names.remove(0)删除第一个元素。输出 names 和 names2。在 names 中验证 Alice B. Toklas 从 names 中移除，而不是从 names2 中移除。

j）创建一个名为 names3 的第三个 ArrayList，使用 Clone 方法实现把 names 复制到 names3.

4.4 Experimental steps（实验步骤）

1.

```
import java.util.Scanner;
```

```
/**
 * 将数组前 n 个值与后面所有值调换顺序
 * @author ******
*/
public class ArrayTest{
    public static void main(String[] args) {
        // TODO Auto-generated method stub
        double[] array=new double[50];
        int length=0;
        int n;
        System.out.print("请输入数组内容：");
        Scanner in=new Scanner(System.in);
        while(in.hasNextDouble())
        {
            array[length]=in.nextDouble();
            length++;
        }
        in.next();
        System.out.print("请输入调换的个数：");
        while(!in.hasNextInt())
        {
            System.out.print("输入错误，请重新输：");
            in.next();
        }
        n=in.nextInt();
        in.close();
        double[] rotation=new double[n];
        for(int i=0;i<n;i++)
        {
            rotation[i]=array[i];
        }
        for(int i=n;i<length;i++)
        {
            array[i-n]=array[i];
        }
        for(int i=0;i<n;i++)
        {
            array[length-n+i]=rotation[i];
        }
        for(int i=0;i<length;i++)
        {
            System.out.print(array[i]+" ");
        }
    }
}
```

2.

```
import java.util.ArrayList;
```

```
/**
 * 创建 ArrayList 并进行一系列的操作及函数调用
 * @author ******
*/
public class ArrayListTest{
    public static void main(String[] args) {
        // TODO Auto-generated method stub
        ArrayList<String> names = new ArrayList<String>();
        System.out.println("初始化: "+names);
        names.add("Alice");
        names.add("Bob");
        names.add("Connie");
        names.add("David");
        names.add("Edward");
        names.add("Fran");
        names.add("Gomez");
        names.add("Harry");
        System.out.println("赋值后: "+names);
        System.out.println("第一个名字: "+names.get(0));
        System.out.println("最后一个名字: "+names.get(names.size()-1));
        System.out.println("数组列表大小为: "+names.size());
        System.out.println("借助 size()输出最后一个名字:
          "+names.get(names.size()-1));
        names.set(names.indexOf("Alice"), "Alice B.Toklas");
        System.out.println("替换后: "+names);
        names.add(names.indexOf("David")+1, "Doug");
        System.out.println("添加后: "+names);
        System.out.print("enhanced for loop 输出: ");
        for(String temp:names)
        {
            System.out.print(" "+temp);
        }
        System.out.println();
        ArrayList<String> names2=new ArrayList<String>();
        names2.addAll(names);
        System.out.println("names2: "+names);
        names.remove(0);
        System.out.println("删除后 names: "+names);
        System.out.println("删除后 names2: "+names2);
        ArrayList<String> names3=(ArrayList<String>) names.clone();
        System.out.println("克隆后 names3: "+names3);
    }
}
```

4.5 Experimental result（实验结果）

实验结果如图 4-4 和图 4-5 所示。

```
请输入数组内容：1 2 3 4 5 6 7 .
请输入数调换的个数：3
4.0 5.0 6.0 7.0 1.0 2.0 3.0
```

Fig4-4. ArrayTest

```
Markers  Properties  Servers  Data Source Explorer  Snippets  Console
<terminated> Lab4_2 [Java Application] C:\Program Files\Java\jre1.8.0_65\bin\javaw.exe (2016年3月31日 上午10:27:25)
初始化：[]
赋值后：[Alice, Bob, Connie, David, Edward, Fran, Gomez, Harry]
第一个名字：Alice
最后一个名字：Harry
数组列表大小为：8
借助size（）输出最后一个名字：Harry
替换后：[Alice B.Toklas, Bob, Connie, David, Edward, Fran, Gomez, Harry]
添加后：[Alice B.Toklas, Bob, Connie, David, Doug, Edward, Fran, Gomez, Harry]
enhanced for loop 输出：Alice B.Toklas Bob Connie David Doug Edward Fran Gomez Harry
names2：[Alice B.Toklas, Bob, Connie, David, Doug, Edward, Fran, Gomez, Harry]
删除后names：[Bob, Connie, David, Doug, Edward, Fran, Gomez, Harry]
删除后names2：[Alice B.Toklas, Bob, Connie, David, Doug, Edward, Fran, Gomez, Harry]
克隆后names3：[Bob, Connie, David, Doug, Edward, Fran, Gomez, Harry]
```

Fig4-5. ArrayListTest

第 5 章　Input Output and Exception Handling（输入输出与异常处理）

5.1　Key points of this chapter（本章要点）

1. Reading and Writing Text Files（文件的读写）

The Scanner class can be used to read text files; The PrintWriter class will be used to write text files

Scanner 类可以用来读取文本文件；PrintWriter 类可以用来写入文本文件。

```
File inputFile = new File("input.txt");
Scanner in = new Scanner(inputFile);
while (in.hasNextLine())
{
  String line = in.nextLine();
  // Process line;
}
in.close();
PrintWriter out = new PrintWriter("output.txt");
out.println("Hello, World!");
out.printf("Total: %8.2f\n", totalPrice);
out.close();
```

2. Text Input and Output（文本的输入和输出）

1）Reading Words（读取单词）

The next method of the Scanner class reads the next string. Consider the loop

接下来的方法展示了使用 Scanner 类中的方法读取一个字符串，考虑使用循环来实现。

```
while (in.hasNext())
{
    String input = in.next();
    System.out.println(input);
}
```

If the user provides the input:
Mary had a little lamb

this loop prints each word on a separate line:

如果用户输入了以下字符串：

Mary had a little lamb

这个循环会以每个单词独占一行的格式输出。

```
Mary
had
a
little
lamb
```

2）Reading Characters（读取字符）

```
Scanner in = new Scanner(……);
in.useDelimiter("");
```

Now each call to next returns a string consisting of a single character.

现在每次使用 next 方法都会返回一个单个字符组成的字符串。

Here is how you can process the characters:

这里的代码提供了如何处理这些字符串：

```
while (in.hasNext())
{
    char ch = in.next().charAt(0);
    Process ch.
}
```

3）Reading Lines（行的读取）

Read each input line into a string:

将输入的每一行读取为一个字符串：

```
while (in.hasNextLine())
{
    String line = nextLine();
    Process line.
}
```

4）Converting Strings to Numbers（字符串类型转换成 number 类型）

```
int populationValue = Integer.parseInt(population);
// suppose that the string is the character sequence
//"303824646". populationValue is the integer 303824646
double price = Double.parseDouble(input);
// suppose input is the string "3.95". price is the floating-point number
//3.95
```

3. Command Line Arguments（命令行参数）

For example, if you start a program with the command line

java ProgramClass -v input.dat

then the program receives two command line arguments: the strings "-v" and "input.dat". It is entirely up to the program what to do with these strings. It is customary to interpret strings starting with a hyphen (-) as program options.

Your program receives its command line arguments in the args parameter of the main method:

举例来说，如果你通过命令行 java ProgramClass -v input.dat 启动一个程序，这个程序会接收两个命令行参数：字符串“-v”和“input dat”。这完全取决于程序如何处理这些字符串，通常程序会将以连字符（-）开头的字符串解释为程序选项。

程序通过 main 方法中的 args 参数来设置命令行参数：

```
public static void main(String[] args)
In our example, args is an array of length 2, containing the strings
args[0]: "-v"
args[1]: "input.dat"
```

4．Exception Handling（异常处理）

There are two aspects to dealing with program errors: detection and handling. In Java, exception handling provides a flexible mechanism for passing control from the point of error detection to a handler that can deal with the error.

处理程序错误一般有两个方面：检测和处理。在 Java 中，异常处理提供了一种灵活的机制，该机制用来控制将异常从检测到错误的地方传递到可以处理错误的处理程序。

1） Throwing an Exception（抛出异常）

```
throw exceptionObject;
```

抛出异常如图 5-1 所示。

```
if (amount > balance)
{
   throw new IllegalArgumentException("Amount exceeds balance");
}
balance = balance - amount;
```

A new exception object is constructed, then thrown.

Most exception objects can be constructed with an error message.

This line is not executed when the exception is thrown.

Fig5-1.　Throwing an Exception

2） Catching Exceptions（捕捉异常）

```
try
{
    statement
    statement
    . . .
}
```

```
catch (ExceptionClass exceptionObject)
{
    statement
    statement
    . . .
}
```

捕捉异常如图 5-2 所示。

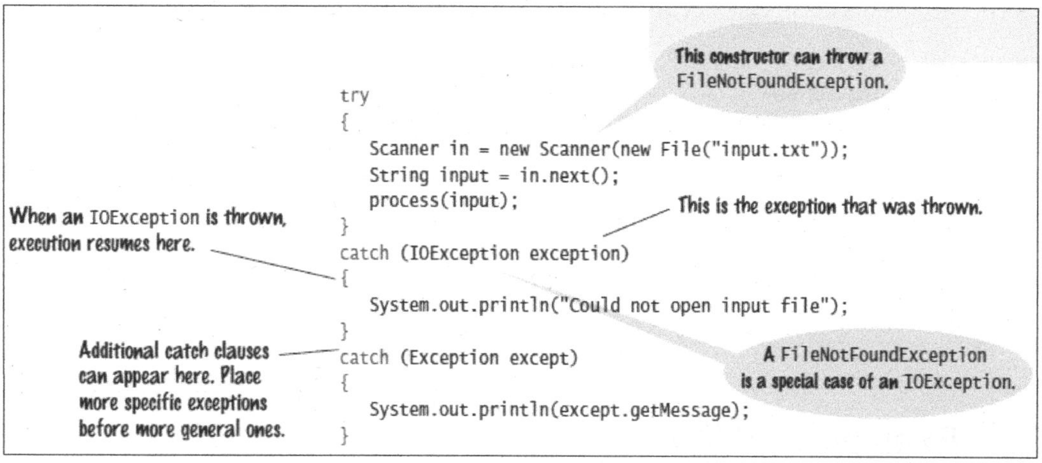

```
try
{
   Scanner in = new Scanner(new File("input.txt"));
   String input = in.next();
   process(input);
}
catch (IOException exception)
{
   System.out.println("Could not open input file");
}
catch (Exception except)
{
   System.out.println(except.getMessage);
}
```

This constructor can throw a FileNotFoundException.
This is the exception that was thrown.
When an IOException is thrown, execution resumes here.
Additional catch clauses can appear here. Place more specific exceptions before more general ones.
A FileNotFoundException is a special case of an IOException.

Fig5-2.　Catching Exception

3）	The throws Clause（throws 语句）

```
modifiers returnType methodName(parameterType parameterName, . . .)
throws ExceptionClass, ExceptionClass, . . .
```

throws 语句如图 5-3 所示。

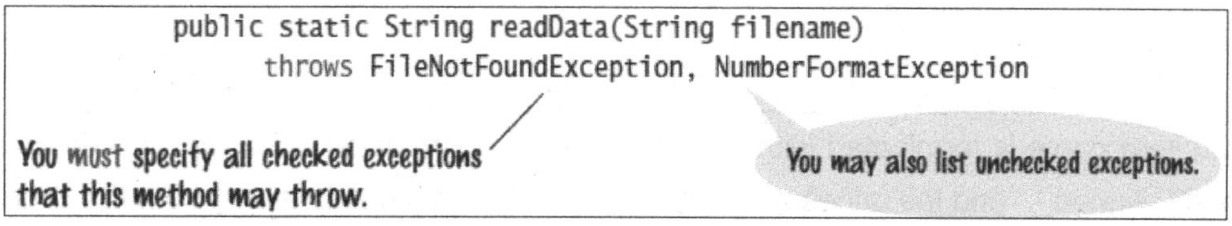

```
public static String readData(String filename)
      throws FileNotFoundException, NumberFormatException
```

You must specify all checked exceptions that this method may throw.
You may also list unchecked exceptions.

Fig5-3.　The throws Clause

4）	The finally Clause（finally 语句）
	Occasionally, you need to take some action whether or not an exception is thrown. The finally construct is used to handle this situation.
	有时候，无论你是否抛出异常，都需要处理一些情况，finally 结构语句就是用来处理这种情况的。

```
try
{
    statement
    statement
    . . .
}
finally
```

```
{
    statement
    statement
    . . .
}
```

finally 语句如图 5-4 所示。

This variable must be declared outside the try block so that the finally clause can access it.
This code may throw exceptions.
This code is always executed, even if an exception occurs.

```
PrintWriter out = new PrintWriter(filename);
try
{
   writeData(out);
}
finally
{
   out.close();
}
```

Fig5-4.　The finally Clause

5.2　Example（例题）

1.	This program reads a file whose lines contain items and prices, like this: 该程序读取一个每行包含某项东西和这件东西价格的文件，如下所示：

```
item name 1: price 1
item name 2: price 2
...
```

Each item name is terminated with a colon. The program writes a file in which the items are left-aligned　and the prices are right-aligned. The last line has the total of the prices. The contents of Input.txt file like this:

每一项物品的名称由冒号结尾。这个程序写入了一个物品左对齐格式和价格右对齐格式的文件。最后一行是文件中每项物品价格的总和。Input.txt 文件的内容如下所示：

```
Price List
Toilet paper: 2.29
Mop: 7.50
Scouring pads: 5
import java.io.File;
import java.io.FileNotFoundException;
import java.io.PrintWriter;
import java.util.Scanner;
public class Items
{
  public static void main(String[] args) throws FileNotFoundException
```

```
    {
      // Prompt for the input and output file names
      Scanner console = new Scanner(System.in);
      System.out.print("Input file: ");
      String inputFileName = console.next();
      System.out.print("Output file: ");
      String outputFileName = console.next();
      // Construct the Scanner and PrintWriter objects for reading and
         writing
      File inputFile = new File(inputFileName);
      Scanner in = new Scanner(inputFile);
      PrintWriter out = new PrintWriter(outputFileName);
      // Read the input and write the output
      double total = 0;
      // We read a line at a time since there may be spaces in the item
        names
      while (in.hasNextLine())
      {
        String line = in.nextLine();
        boolean found = false;
        String item = "";
        double price = 0;
        for (int i = 0; !found && i < line.length(); i++)
        {
          char ch = line.charAt(i);
          if (ch == ':')
          {
            found = true;
            item = line.substring(0, i + 1);
            price = Double.parseDouble(line.substring(i + 1).trim());
            total = total + price;
          }
        }
        // If no colon was found, we skip the line
        if (found)
        {
          out.printf("%-20s%10.2f\n", item, price);
        }
      }
      out.printf("%-20s%10.2f\n", "Total:", total);
      in.close();
      out.close();
    }
  }
```

在运行此程序时输入数据如图 5-5 所示。

```
Input file: d:\\input.txt
Output file: d:\\output.txt
```

Fig5-5. Input the data When you run this program

When it has been finished, you will get the contents of the output.txt.

当程序运行完成时，你会得到 output.txt 文件的内容如下所示：

```
Toilet paper:          2.29
Mop:                   7.50
Scouring pads:         5.00
Total:                14.79
```

2. the main method of the Data Analyzer program, it catches the exceptions, prints appropriate error messages.

main 方法主要实现了数据分析器的功能，它捕获异常，输出适当的错误消息。

```
import java.util.InputMismatchException;
import java.util.NoSuchElementException;
import java.util.Scanner;
public class ExceptionDemo
{
   public static void main(String[] args)
   {
      String inputValues = "two 42 43 three 44 ";
      // This scanner reads the values from the given string.
      Scanner in = new Scanner(inputValues);
      boolean done = false;
      while (!done)
      {
         try
         {
            int n = in.nextInt();
            System.out.println("Read integer " + n);
            if (n == 42) { throw new IllegalArgumentException("Ugh! 42"); }
            String str = in.next();
            System.out.println("Read string " + str);
            n = Integer.parseInt(str);
            System.out.println("Parsed integer " + n);
         }
         catch (NumberFormatException ex)
         {
            System.out.println(ex.getMessage());
         }
         catch (IllegalArgumentException ex)
         {
            System.out.println(ex.getMessage());
         }
         catch (InputMismatchException ex)
         {
            // We "fix" the problem by removing the offending input
            System.out.println("Removing " + in.next());
         }
```

```
            catch (NoSuchElementException ex) // This happens at the end of
the input
            {
                ex.printStackTrace();
                System.out.println("Terminating the loop.");
                done = true;
            }
        }
    }
}
```

5.3　Experimental contents（实验内容）

1. Write a program to store multiple memos in a file. Allow a user to enter a topic and the text of a memo (the text of the memo is stored as a single line and thus cannot contain a return character). Store the topic, the date stamp of the memo, and the memo message.

   编写一个程序将多条备忘录存储在一个文件中。允许用户输入备忘录的主题和备忘录的内容（每条备忘录存储为一行，所以内容不能包含换行字符），可以存储备忘录的主题、时间和备忘录的内容。

Creating a java.util.Date object with no arguments will initialize the Date object to the current time and date. A date stamp is obtained by calling the Date.toString() method.

创建没有参数的 java.util.Date 对象会将 Date 对象初始化为当前时间和日期。通过调用 Date.toString()方法获取时间戳。

Use a text editor to view the contents of the output and to check that the information is stored correctly.

使用一个文本编辑器查看输出的内容，并检查信息是否正确存储。

2. Write a program that processes command line arguments. The arguments are a mixture of numbers (ints and doubles). Concatenate all the arguments into a single string. Scan the string, then print each number on a separate line and whether it is an int or a double.

   编写一个处理命令行参数的程序。参数包含 int 型和 double 型的数据。将所有参数合并为一个字符串，然后扫描这个字符串，将整型数据以一行输出，浮点型数字为一行输出。

Test your program with each of the following lists of arguments:

使用下面的每一个参数列表测试您的程序：

```
a) 1 2 3 4 5
b) 1.1  2.2  3.3  4.4
```

```
c) 1  2.9  3  4.9  5  6.9
```

3.	Start with the code below and complete the getInt method. The method should prompt the user to enter an integer. Scan the input the user types. If the input is not an int, throw an IOException; otherwise, return the int. Modify the main program so that it catches and prints the IOException.
	继续编写下面给出的代码，完成 getInt 方法。这个方法提示用户输入整数，扫描用户输入的数据类型，如果输入的不是整数，抛出一个 IOException，否则，返回输入的整数。修改主函数，使主函数可以捕捉 IOException，并输出这个 IOException。

```
import java.util.Scanner;
public class Throwing
{
   public static void main(String[] args)
   {
      int x = getInt();
      System.out.println(x);
   }
   public static int getInt()
   {
      // your code goes here
   }
}
```

5.4 Experimental steps（实验步骤）

1.

```
import java.io.FileWriter;
import java.io.IOException;
import java.text.DateFormat;
import java.text.SimpleDateFormat;
import java.util.Date;
import java.util.Scanner;
/**
 * 备忘录的生成
 * @author *****
*/
public class Memo {
    public static void main(String[] args){
        // TODO Auto-generated method stub
        Scanner in=new Scanner(System.in);
        Date date=new Date();
```

```
        //String time=date.toString();
        DateFormat format=new SimpleDateFormat("yyyy-MM-dd HH:mm:ss");
        String time=format.format(date);
        try
        {
            FileWriter pw = new FileWriter("E:/test/memo.txt",true);
            System.out.print("请输入备忘录标题：");
            String str1=in.next();
            System.out.print("请输入备忘录内容：");
            String str2=in.next();
            pw.write(time);
            pw.write("   "+str1+"   "+str2);
            pw.write("\r\n");
            pw.close();
        }
        catch (IOException e)
        {
            // TODO Auto-generated catch block
            e.printStackTrace();
        }
        finally
        {
            in.close();
        }
    }
}
```

2.

```
import java.util.Scanner;
/**
 * 从字符串中分离整数和浮点数
 * @author ******
*/
public class CommandLine {

    public static void main(String[] args) {
        // TODO Auto-generated method stub
        String[] str2=new String[20];
        int[] array1=new int[20];           //存储 Integer 型数值
        double[] array2=new double[20];     //存储 double 型数值
        int t1=0;                           //记录 Integer 型个数
        int t2=0;                           //记录 double 型个数
        str2=args;
        for(int i=0;i<str2.length;i++)
        {
```

```
            boolean hasDot=false;
            for(int j=0;j<str2[i].length();j++)
            {
                if(str2[i].charAt(j)=='.')
                {
                    hasDot=true;
                }
            }
            if(hasDot)
            {
                array2[t2++]=Double.parseDouble(str2[i]);
            }
            else
            {
                array1[t1++]=Integer.parseInt(str2[i]);
            }
        }
        if(t1==0)
        {
            System.out.println("没有整型");
        }
        else
        {
            System.out.println("整型有：");
            for(int i=0;i<t1;i++)
            {
                System.out.print(array1[i]+"  ");
            }
            System.out.println();
        }
        if(t2==0)
        {
            System.out.println("没有浮点型");
        }
        else
        {
            System.out.print("浮点型有：");
            for(int i=0;i<t2;i++)
            {
                System.out.print(array2[i]+"  ");
            }
        }
    }
}
```

3.

```
import java.io.IOException;
import java.util.Scanner;
/**
 * 异常处理
 * @author *******
*/
public class ExceptionHandle{
    public static void main(String[] args){
        try
        {
            int x = getInt();
            System.out.println(x);
        }
        catch(IOException e)
        {
            e.printStackTrace();
            System.out.println("errors!");
        }
    }
    public static int getInt () throws IOException
    {
        Scanner in=new Scanner(System.in);
        if(!in.hasNextInt())
        {
            in.close();
            throw new IOException();
        }
        int temp=in.nextInt();
        in.close();
        return temp;
    }
}
```

5.5　Experimental result（实验结果）

实验结果如图 5-6～图 5-10 所示。

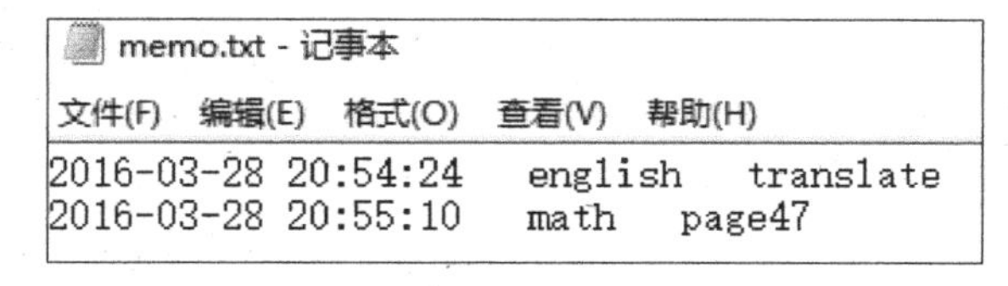

Fig5-6.　the Result of Memo

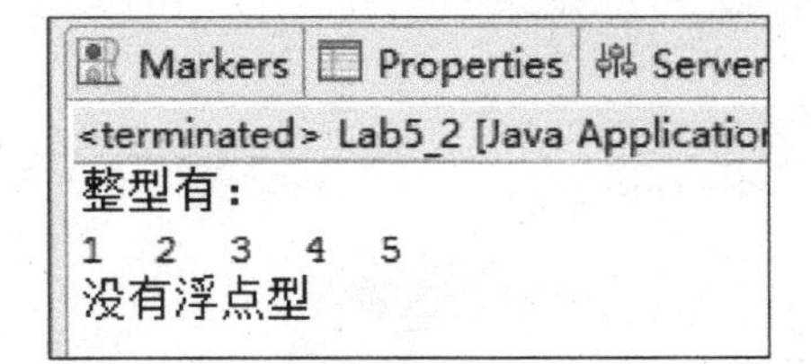

Fig5-7. Case 1 of the Result of CommandLine

```
Markers  Properties  Servers
<terminated> Lab5_2 [Java Application] C:\Pr
没有整型
浮点型有：1.1  2.2  3.3  4.4
```

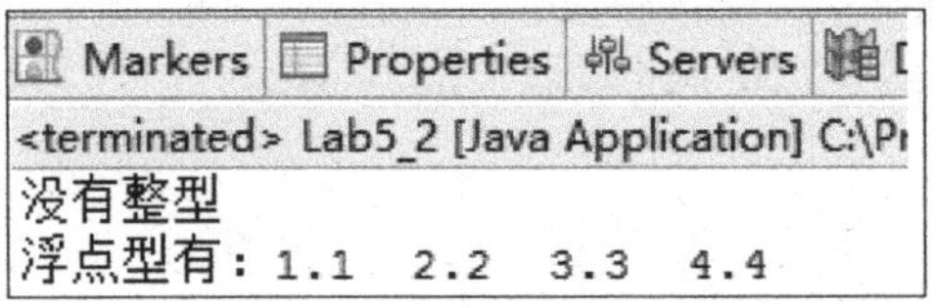

Fig5-8. Case 2 of the Result of CommandLine

```
整型有:
1  3  5
浮点型有: 2.9  4.9  6.9
```

Fig5-9. Case 3 of the Result of CommandLine

```
1.1
java.io.IOExceptionerrors!

        at ExceptionHandle.getInt(ExceptionHandle.java:26)
        at ExceptionHandle.main(ExceptionHandle.java:11)
```

Fig5-10. The Result of ExceptionHandle

第 6 章　Objects and Classes（类与对象）

6.1　Key points of this chapter（本章要点）

1. Diagram of a Class（类的图表表示）

类的图表表示如图 6-1 所示。

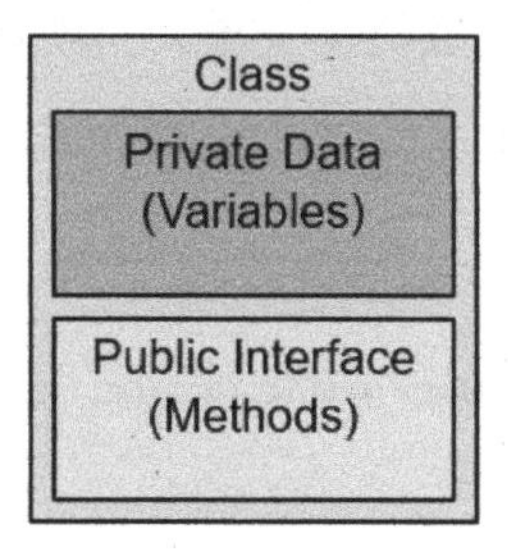

Fig6-1.　Diagram of a Class

2. Implementing a Class（类的实现）

```
public class ClassName
{
   private typeName variableName;
   . . .
}
```

实例化变量声明如图 6-2 所示。

```
public class Counter
{
   private int value;
   . . .
}
```

Instance variables should always be private.

Each object of this class has a separate copy of this instance variable.

Type of the variable

Fig6-2.　Instance Variable Declaration（实例化变量声明）

3. Implementing Instance Methods（实例化方法实现）

```
modifiers returnType methodName(parameterType parameterName, . . . )
{
   method body
}
```

实例化方法如图 6-3 所示。

```
public class CashRegister
{
   . . .
   public void addItem(double price)
   {
      itemCount++;
      totalPrice = totalPrice + price;
   }
   . . .
}
```

Explicit parameter

Instance variables of the implicit parameter

Fig6-3. Instance Methods（实例化方法）

3. Constructor（构造函数）

A constructor initializes the instance variables of an object. The constructor is automatically called whenever an object is created with the new operator. 构造函数会初始化一个对象的实例化变量。当创建一个带有 new 运算符对象时，会自动调用构造函数。

构造函数如图 6-4 所示。

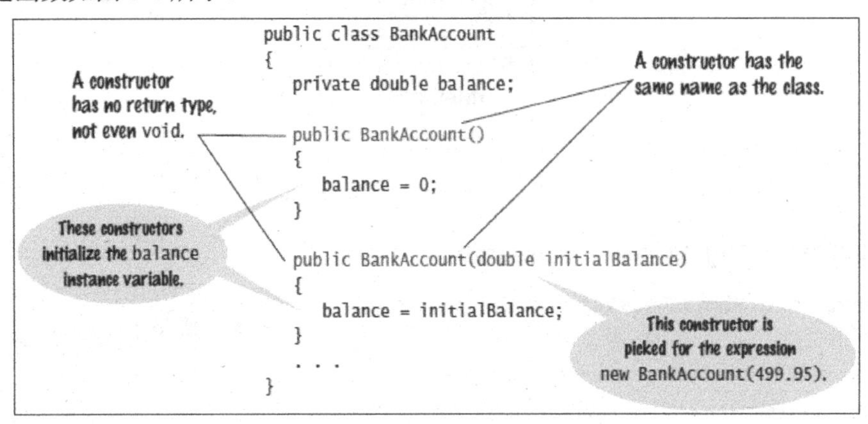

Fig6-4. Constructor

4. Object References（对象的引用）

We use the technical term object reference to denote the memory location of an object. 我们用一个对象的引用来指示这个对象在内存中的地址。

1）	Shared References（共享引用）
	You can have two (or more) object variables that store references to the same object, for example by assigning one to the other. 你可以为同一个对象创建两个或者多个引用对象变量，例如将一个指向另一个。

```
CashRegister reg2 = reg1;
```

2）	The null Reference（空引用）
	An object reference can have the special value null if it refers to no object at all. It is common to use the null value to indicate that a value has never been set. For

example,

如果一个引用对象没有指向任何一个对象，那它就为空。用一个 null 去标明一个没有设置的值是很常见的做法。例如，

```
String middleInitial = null; // No middle initial
```

3） The this Reference（this 引用）

Every instance method receives the implicit parameter in a variable called this.

每个实例化对象方法会接收一个名为 this 的隐含参数。

5. Static Variables and Methods(静态变量和方法)

Sometimes, a value properly belongs to a class, not to any object of the class. You use a static variable for this purpose. Sometimes a class defines methods that are not invoked on an object. Such a method is called a static method.

有时候，一个值可能是属于类，而不属于类的对象，静态变量就可以达到这一目的。也有时候一个类定义的方法并不是给对象调用，这样的方法叫做静态方法。

6.2 Example（例题）

Let's consider an example. We want to use objects that simulate cash registers. A cashier who rings up a sale presses a key to start the sale, then rings up each item. A display shows the amount owed as well as the total number of items purchased. In our simulation, we want to call the following methods on a cash register object:

1) Add the price of an item.

2) Get the total amount owed, and the count of items purchased.

3) Clear the cash register to start a new sale.

让我们来考虑以下例子，我们想用一个对象模仿现金收银。收银员按下销售键开始，然后逐个输入款项，显示器显示应付款和所有购买东西的总价。在我们的模拟中，想用现金收银对象调用以下方法：

1）计算所有价格的总和。

2）得到总价和购买物品的具体明细。

3）清空现金收银数据，开始新的销售记录。

```
/**
  A simulated cash register that tracks the item count and
  the total amount due.
 */
public class CashRegister
{
  private int itemCount;
  private double totalPrice;
```

```
   /**
      Constructs a cash register with cleared item count and total.
   */
   public CashRegister()
   {
      itemCount = 0;
      totalPrice = 0;
   }
   /**
      Adds an item to this cash register.
      @param price the price of this item
   */
   public void addItem(double price)
   {
      itemCount++;
      totalPrice = totalPrice + price;
   }
   /**
      Gets the price of all items in the current sale.
      @return the total amount
   */
   public double getTotal()
   {
      return totalPrice;
   }
     /**
      Gets the number of items in the current sale.
      @return the item count
   */
   public int getCount()
   {
      return itemCount;
   }
   /**
      Clears the item count and the total.
   */
   public void clear()
   {
      itemCount = 0;
      totalPrice = 0;
   }
}
/**
   This program tests the CashRegister class.
*/
public class CashRegisterTester
{
```

```
    public static void main(String[] args)
    {
        CashRegister register1 = new CashRegister();
        register1.addItem(1.95);
        register1.addItem(0.95);
        register1.addItem(2.50);
        System.out.println(register1.getCount());
        System.out.println("Expected: 3");
        System.out.printf("%.2f\n", register1.getTotal());
        System.out.println("Expected: 5.40");
    }
}
```

6.3 Experimental contents（实验内容）

<table>
<tr><td rowspan="2">1.</td><td>Design a class Message that models an e-mail message. A message has a recipient, a sender, and a message text. Support the following methods:
1) A constructor that takes the sender and recipient.
2) A method append that appends a line of text to the message body.
3) A method toString that makes the message into one long string like this: "From: Harry Morgan\nTo: Rudolf Reindeer\n . . .".
Write a program that uses this class to make a message and print it.</td></tr>
<tr><td>模仿邮箱设计一个 Message 类，有收件人、发件人和邮件，并用到以下方法：
1）带有发件人和收件人参数的构造函数。
2）添加邮件内容的方法。
3）使用 toString 方法将邮件信息相连接，例如："From: Harry Morgan\nTo: Rudolf Reindeer\n . . ."。
编写一个程序，用这个类创建一封邮件并把它输出来。</td></tr>
</table>

<table>
<tr><td rowspan="2">2.</td><td>Design a class Mailbox that stores e-mail messages, using the Message class of Exercise 1. Implement the following methods:
1) public void addMessage(Message m)
2) public Message getMessage(int i)
3) public void removeMessage(int i)</td></tr>
<tr><td>利用练习 1 中的 Message 对象，设计一个 Mailbox 对象来存储邮件。实现以下方法：
1) public void addMessage(Message m)
2) public Message getMessage(int i)
3) public void removeMessage(int i)</td></tr>
</table>

6.4 Experimental steps（实验步骤）

1.

```
/**
 * 邮件类
 * @author ******
 *
 */
public class Message {
    protected String recipient;
    protected String sender;
    protected String messageText;
    /**
     * 构造函数
     * @param recipient
     * @param sender
     */
    public Message(String recipient, String sender) {
        super();
        this.recipient = recipient;
        this.sender = sender;
        messageText="";
    }
    /**
     * 向邮件文本中加入一行内容
     * @param text
     */
    public void append(String text){
        messageText=messageText+text+"\n";
    }
    /**
     * 将邮件按输出格式标准化
     */
    @Override
    public String toString() {
        return "From:"+recipient+"\n"+"to:"+sender+"\n"+messageText;
    }
}
```

2.

```
import java.util.ArrayList;
/**
 * 邮箱
 * @author *******
 *
 */
public class Mailbox {
```

```
protected ArrayList<Message> mail;
public int count=0;
/**
 * 构造函数
 */
public Mailbox() {
    super();
    mail=new ArrayList<Message>();
    count=0;
}
/**
 * 增加邮件
 * @param m
 */
public void addMessage(Message m)
{
    mail.add(m);
    count++;
}
/**
 * 提取邮件
 * @param i
 * @return
 */
public Message getMessage(int i)
{
    if(i>mail.size())
    {
        System.out.println("errors!");
        return null;
    }
    return mail.get(i);
}
/**
 * 删除邮件
 * @param i
 */
public void removeMessage(int i)
{
    if(i>mail.size())
    {
        System.out.println("errors!");
    }
    else
    {
        mail.remove(i);
        count--;
```

```
        }
    }
}
```

3.	write a test class to use the designed class. 编写一个测试类完成以上类的使用。

```
import java.util.Scanner;
import org.junit.Test;
public class test1 {
    @Test
    public void test() {
        //fail("Not yet implemented");
        String recipient=new String();
        String sender=new String();
        String messageText=new String();
        Scanner in=new Scanner(System.in);
        System.out.print("请输入发件人姓名：");
        recipient=in.next();
        System.out.print("请输入收件人姓名：");
        sender=in.next();
        Message email=new Message(recipient, sender);
        System.out.println("请输入邮件内容：");
        boolean isEnd=false;
        do{
            messageText=in.next();
            email.append(messageText);
            if(messageText.charAt(messageText.length()-1)=='#')
            {
                isEnd=true;
            }
        }while(!isEnd);
        in.close();
        System.out.println(email.toString());

    }
    @Test
    public void test2()
    {
        Mailbox box=new Mailbox();
        Message mail1=new Message("Mr.Li","Mr.Luo");
        mail1.append("hello,1");
        Message mail2=new Message("Mr.Li","Mr.Luo");
        mail2.append("hello,2");
        Message mail3=new Message("Mr.Li","Mr.Luo");
        mail3.append("hello,3");
        box.addMessage(mail1);
        box.addMessage(mail2);
```

```
        box.addMessage(mail3);
        for(int i=0;i<box.count;i++)
        {
            System.out.println(box.getMessage(i).toString());
        }
        box.removeMessage(1);
        for(int i=0;i<box.count;i++)
        {
            System.out.println(box.getMessage(i).toString());
        }
    }
}
```

6.5 Experimental result（实验结果）

实验结果如图 6-5 和 6-6 所示。

```
<terminated> test1.test [JUnit] C:\Progr
请输入发件人姓名: Mr.Li
请输入收件人姓名: Mr.Liu
请输入邮件内容:
Hello
I'm Mr.Li#
From:Mr.Li
to:Mr.Liu
Hello
I'm
Mr.Li#
```

Fig6-5　Test Message Class

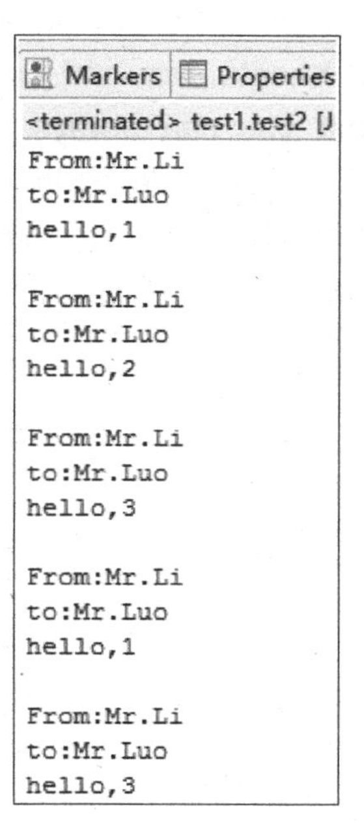

Fig6-6　Test Mailbox Class

第 7 章 Inheritance and Interfaces（继承与接口）

7.1 Key points of this chapter（本章要点）

1. Inheritance Hierarchies（继承关系）

In object-oriented design, inheritance is a relationship between a more general class (called the superclass) and a more specialized class (called the subclass). The subclass inherits data and behavior from the superclass. The substitution principle states that you can always use a subclass object when a superclass object is expected. For example, an inheritance hierarchy for these question types.

在面向对象编程中，继承是用来表示基类（也称为超类）和派生类（也称为子类）之间的关系。派生类从基类那里继承基类所拥有的数据和行为。你可以用派生类的对象去替代基类的对象。图 7-1 是一个 Question 类的继承关系图。

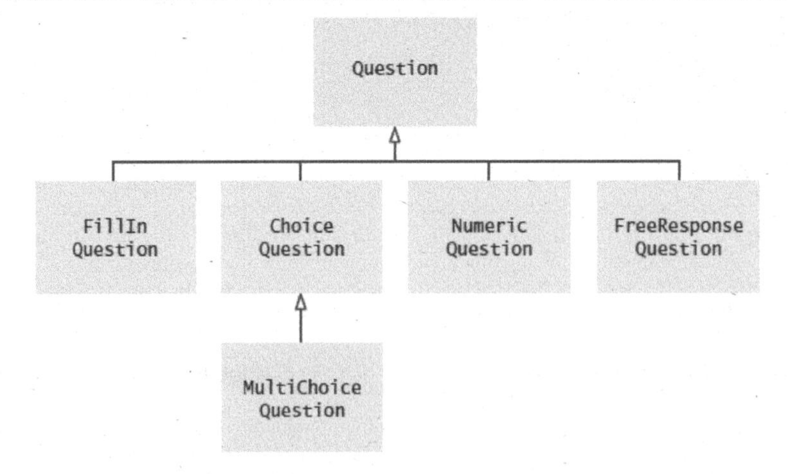

Fig7-1. Inheritance Hierarchy of Question Types

```
public class Question
{
   private String text;
   private String answer;
   public void setText(String questionText)    // Sets the question text.
   { . . . }
```

```
        public void setAnswer(String correctResponse) // Sets the answer for this question.
          { . . . }
        public boolean checkAnswer(String response) // Checks a given response for correctness.
          { . . . }
          public void display() //Displays this question.
          { . . . }
    }
```

2. Implementing Subclasses（实现派生类）

In Java, you form a subclass by specifying what makes the subclass different from its superclass. Subclass objects automatically have the instance variables that are declared in the superclass. You only declare instance variables that are not part of the superclass objects.

在 Java 中，你可以指定派生类中与基类中不同的东西。派生类对象中已经包含了基类中声明了的实例变量。使用时你只需要声明不同于基类的变量即可。

```
    public class SubclassName extends SuperclassName
    {
        instance variables
        methods
    }
```

类的选择如图 7-2 所示。

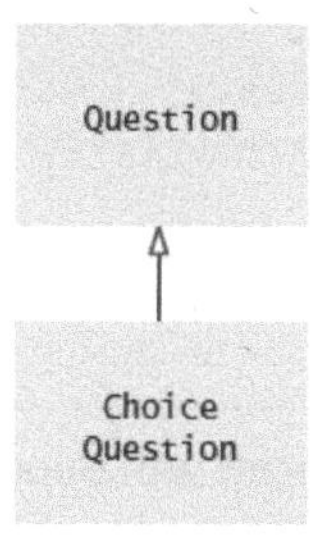

Fig7-2　The ChoiceQuestion Class

子类声明如图 7-3 所示。

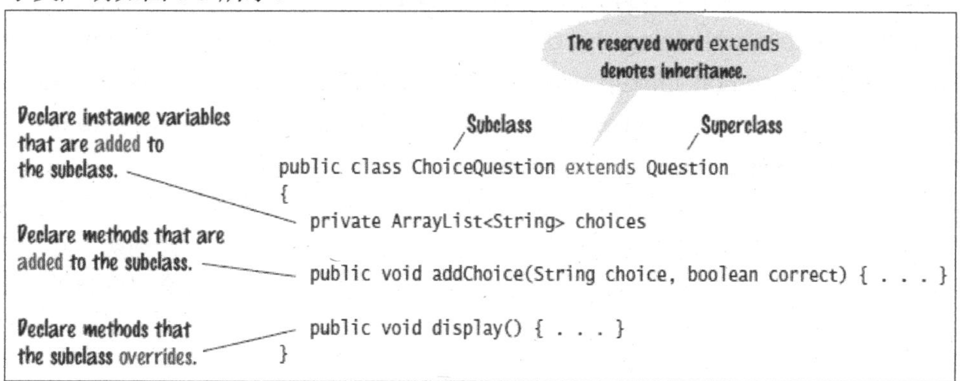

Fig7-3　Subclass Declaration

3. Overriding Methods（重写方法）

The subclass inherits the methods from the superclass. If you are not satisfied with the behavior of an inherited method, you override it by specifying a new implementation in the subclass. Instead, you can call the same method of the superclass, by using the reserved word super.

派生类已经继承了基类的方法。如果你不满足于继承来的方法，可以通过 override（重写）的方法来指定新的方法实现。你也可以使用 super 关键字来调用基类中命名相同的方法。

In order to specify another constructor, you use the super reserved word, together with the arguments of the superclass constructor, as the first statement of the subclass constructor.

在声明构造函数时，可以用 super 关键字来向基类的构造函数传递数据，就像下面这个派生类的构造函数中的第一行代码所展示的内容。

```
public ClassName(parameterType parameterName, . . .)
{
    super(arguments);
    . . .
}
```

父类的构造函数与初始化如图 7-4 所示。

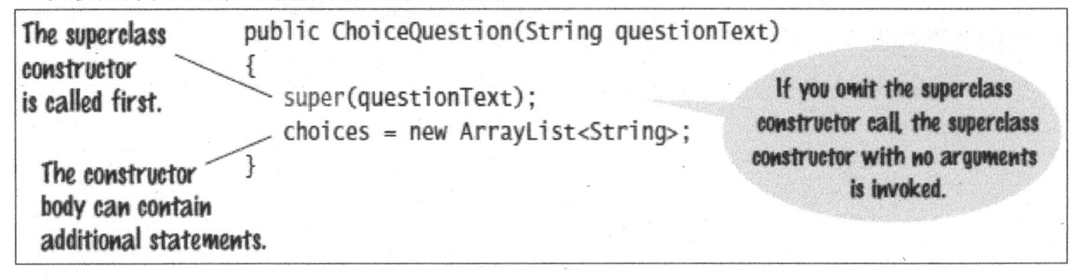

Fig7-4 Constructor with Superclass Initializer

4. Polymorphism（多态性）

In Java, method calls are always determined by the type of the actual object, not the type of the variable containing the object reference. This is called dynamic method lookup. Dynamic method lookup allows us to treat objects of different classes in a uniform way. This feature is called polymorphism. We ask multiple objects to carry out a task, and each object does so in its own way.

在 Java 中，函数的调用由对象的实际类型决定，而不是由对象的引用类型决定的。这被称为动态方法查询。动态方法查询允许我们以相同的方式对待不同的类。这个特征就被称为多态性。我们要求许多的对象执行同一个任务，每个对象都会以各自的方式完成任务。

Therefore, we can simply declare the parameter variable of the presentQuestion method to have the type Question:

因此，我们可以简单地声明 presentQuestion 方法的参数变量类型

```
public static void presentQuestion(Question q)
{
    q.display();
    System.out.print("Your answer: ");
    Scanner in = new Scanner(System.in);
    String response = in.nextLine();
    System.out.println(q.checkAnswer(response));
}
Question first = new Question();
. . .
presentQuestion(first);// OK to pass a Question
ChoiceQuestion second = new ChoiceQuestion();
. . .
presentQuestion(second); // OK to pass a ChoiceQuestion
Question first = new Question();
```

5. Abstract Classes（抽象类）

An abstract class is a class that cannot be instantiated. An abstract method is a method whose implementation is not specified.

抽象类是一种不能创建实例的类。抽象方法是一种没有指定实现内容的方法。

```
public abstract class Account
{
    public abstract void deductFees();
    . . .
}
public class SavingsAccount extends Account // Not abstract
{
    . . .
    public void deductFees() // Provides an implementation
    {
    . . .
    }
}
```

6. Final Methods and Classes（最终的方法与类）

Occasionally, you may want to do the opposite and prevent other programmers from creating subclasses or from overriding certain methods. In these situations, you use the final reserved word. For example, the String class in the standard Java library has been declared as

有的时候，你可能想要阻止其他程序创建子类或重写方法。基于这种情况，你可以使用 final 关键字。例如，String 类在 Java 规范中被声明为如下：

```
public final class String { . . . }
```

That means that nobody can extend the String class. When you have a reference of type String, it must contain a String object, never an object of a subclass. You can also declare individual methods as final:

这就意味着，没人能继承 String 类。当你使用到 String 类型时，只能使用 String 对象，不能使用该对象的子类。你也能在方法声明中使用 final 关键字：

```
public final class String { . . . }
public class SecureAccount extends BankAccount
{
    . . .
    public final boolean checkPassword(String password)
    {
    . . .
    }
}
```

7. Developing an Inheritance Hierarchy（如何开发一个继承关系的类）

Step 1 List the classes that are part of the hierarchy;
Step 2 Organize the classes into an inheritance hierarchy;
Step 3 Determine the common responsibilities;
Step 4 Decide which methods are overridden in subclasses;
Step 5 Declare the public interface of each subclass;
Step 6 Identify instance variables;
Step 7 Implement constructors and methods;
Step 8 Construct objects of different subclasses and process them.

步骤 1. 列出继承关系中的类；
步骤 2. 组织继承关系中的类；
步骤 3. 确定它们之间的共同责任；
步骤 4. 决定哪些方法要在子类重写；
步骤 5. 声明每个子类的公共接口；
步骤 6. 确认实例变量；
步骤 7. 实现构造函数和方法；
步骤 8. 构造不同的子类对象。

8. Interface Types（接口类型）

```
Declaring: public interface InterfaceName
{
method declarations
}
Implementing: public class ClassName implements InterfaceName,
InterfaceName, . . .
{
    instance variables
    methods
}
```

接口类型如图 7-5 所示。

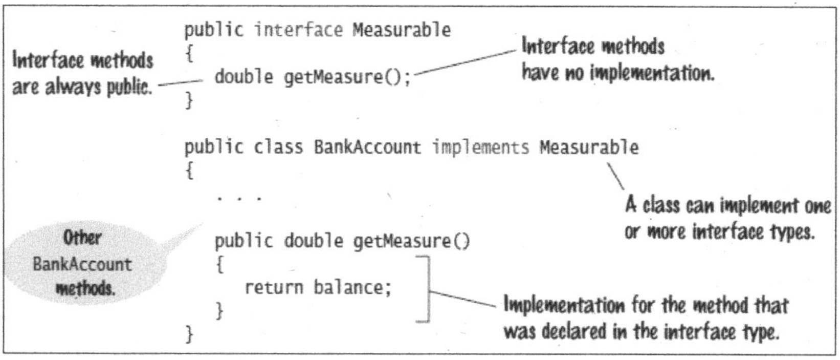

```
public interface Measurable
{
   double getMeasure();
}

public class BankAccount implements Measurable
{
   . . .

   public double getMeasure()
   {
      return balance;
   }
}
```

Interface methods are always public.

Interface methods have no implementation.

A class can implement one or more interface types.

Other BankAccount methods.

Implementation for the method that was declared in the interface type.

Fig7-5　Interface Types

7.2　Example（例题）

You can use inheritance and implement ChoiceQuestion as a subclass of the Question class.

你可以使用继承和实现的方法来创建 ChoiceQuestion 类。

1．Question.java

```
/**
  A question with a text and an answer.
*/
public class Question
{
  private String text;
  private String answer;
  /**
    Constructs a question with empty question and answer.
  */
  public Question()
  {
    text = "";
    answer = "";
  }
  /**
    Sets the question text.
    @param questionText the text of this question
  */
  public void setText(String questionText)
  {
    text = questionText;
  }
  /**
```

```
      Sets the answer for this question.
      @param correctResponse the answer
   */
   public void setAnswer(String correctResponse)
   {
      answer = correctResponse;
   }
   /**
      Checks a given response for correctness.
      @param response the response to check
      @return true if the response was correct, false otherwise
   */
   public boolean checkAnswer(String response)
   {
      return response.equals(answer);
   }
   /**
      Displays this question.
   */
   public void display()
   {
      System.out.println(text);
   }
}
```

2. ChoiceQuestion.java

```
import java.util.ArrayList;
/**
   A question with multiple choices.
*/
public class ChoiceQuestion extends Question
{
   private ArrayList<String> choices;

   /**
      Constructs a choice question with no choices.
   */
   public ChoiceQuestion()
   {
      choices = new ArrayList<String>();
   }
   /**
      Adds an answer choice to this question.
      @param choice the choice to add
      @param correct true if this is the correct choice, false otherwise
   */
   public void addChoice(String choice, boolean correct)
   {
```

```
        choices.add(choice);
        if (correct)
        {
           // Convert choices.size() to string
           String choiceString = "" + choices.size();
           setAnswer(choiceString);
        }
     }
     public void display()
     {
        // Display the question text
        super.display();
        // Display the answer choices
        for (int i = 0; i < choices.size(); i++)
        {
           int choiceNumber = i + 1;
           System.out.println(choiceNumber + ": " + choices.get(i));
        }
     }
  }
```

3. QuestionDemo.java

```
  import java.util.Scanner;
  /**
     This program shows a simple quiz with two question types.
  */
  public class QuestionDemo
  {
     public static void main(String[] args)
     {
        Question first = new Question();
        first.setText("Who was the inventor of Java?");
        first.setAnswer("James Gosling");
        ChoiceQuestion second = new ChoiceQuestion();
        second.setText("In which country was the inventor of Java born?");
        second.addChoice("Australia", false);
        second.addChoice("Canada", true);
        second.addChoice("Denmark", false);
        second.addChoice("United States", false);
        presentQuestion(first);
        presentQuestion(second);
     }
     /**
        Presents a question to the user and checks the response.
        @param q the question
     */
     public static void presentQuestion(Question q)
     {
```

```
        q.display();
        System.out.print("Your answer: ");
        Scanner in = new Scanner(System.in);
        String response = in.nextLine();
        System.out.println(q.checkAnswer(response));
    }
}
```

7.3 Experimental contents（实验内容）

1） Consider using the following Card class.
考虑使用下面的 Card 类。

```
public class Card
{
    private String name;
    public Card()
    {
        name = "";
    }
    public Card(String n)
    {
        name = n;
    }
    public String getName()
    {
        return name;
    }
    public boolean isExpired()
    {
        return false;
    }
    public String format()
    {
        return "Card holder: " + name;
    }
}
```

Use this class as a superclass to implement a hierarchy of related classes:
用这个类作为基类来执行相关类的继承关系结构：

Class	Data
IDCard (superclass: Card)	ID number
CallingCard(superclass: Card)	Card number, PIN
DriverLicense(superclass: CallingCard)	Expiration year

Write declarations for each of the subclasses. For each subclass, supply private instance variables.

为每个派生类写出声明，并提供私有实例变量。

2） Implement constructors for each of the three subclasses. Each constructor should call the superclass constructor to set the name. Here is one example:

实现三个派生类的构造函数。每个构造函数必须调用基类的构造函数去设置名字。例如：

```
public IDCard(String n, String id)
{
   super(n);
   idNumber = id;
}
```

3） Replace the implementation of the format method for the three subclasses. The methods should produce a formatted description of the card details. The subclass methods should call the superclass format method to get the formatted name of the cardholder.

重新实现三个派生类的 format 方法。这些类中的该方法实现 card 类详细信息的格式化说明。派生类方法应该调用基类的 format 方法来获得 formatted 的名字。

4） Implement toString methods for the Card class and its three subclasses. The methods should print:

在 Card 类和三个派生类中重写 toString 方法。输出类的名字，实例变量的值：

```
    the name of the class
    the values of all instance variables (including inherited instance
variables)
```

Typical formats are:

典型的格式：

```
Card[name=Edsger W. Dijkstra]
CallingCard[name=Bjarne Stroustrup][number=4156646425,pin=2234]
Write the code for your toString methods.
```

在 toString()中写出实现的代码。

5） Implement equals methods for the Card class and its three subclasses. Cards are the same if the objects belong to the same class, and if the names and other information (such as the expiration year for driver licenses) match.

在 Card 类和三个派生类中实现 equals 方法。必须是来自同一个类的对象，而且对象的名字属性和其他信息（例如驾驶证的有效期限）也要相同才认为是相同的 Cards。

Give the code for your equals methods.

写出你的 equals()函数实现代码。

6）	Write a test Class to test these classes. 写出一个测试类来测试这些类。

7.4 Experimental steps（实验步骤）

1.

```
/**
 * Card
 * @author ********
 *
 */
public class Card
{
  private String name;
  public Card()
  {
    name = "";
  }
  public Card(String n)
  {
    name = n;
  }
  public String getName()
  {
    return name;
  }
  public boolean isExpired()
  {
    return false;
  }
  public String format()
  {
    return "Card holder: " + name;
  }
  public String toString()
  {
     return "Card[name="+name+"]";
  }
  public boolean equals(Object o)
  {
     if(this==o)
        return true;
     if(o==null)
        return false;
```

```
        if(o.getClass().getName()==this.getClass().getName())
        {
            Card c=(Card)o;
            if(c.name.equals(this.name))
                return true;
        }
        return false;
    }
}
```

2.

```
/**
 * IDCard
 * @author ********
 *
 */
public class IDCard extends Card {
    private String IDNumber;
    public IDCard(String name,String id)
    {
        super(name);
        IDNumber=id;
    }
    public String toString()
    {
        return "IDCard[name="+getName()+",IDNumber="+IDNumber+"]";
    }
    public String format()
    {
        return super.format()+" IDNumber:"+IDNumber;
    }
     public boolean equals(Object o)
        {
            if(this==o)
                return true;
            if(o==null)
                return false;
            if(o.getClass()==this.getClass())
            {
                IDCard c=(IDCard)o;
                if(c.getName().equals(this.getName())
                        &&c.IDNumber==this.IDNumber
                        )
                    return true;
            }
            return false;
        }
}
```

3.

```
/**
 * CallingCard
 * @author ********
 *
 */
public class CallingCard extends Card {
    private String cardNumber;
    private String PIN;
    public CallingCard(String name,String num,String p)
    {
        super(name);
        cardNumber=num;
        PIN=p;
    }
    public String format()
    {
        return super.format()+" cardNumber:"+cardNumber+" PIN:"+PIN;
    }
    public String getPIN()
    {
        return PIN;
    }
    public String getCardNumber()
    {
        return cardNumber;
    }
     public boolean equals(Object o)
       {
          if(this==o)
             return true;
          if(o==null)
             return false;
          if(o.getClass()==this.getClass())
          {
             CallingCard c=(CallingCard)o;
             if(c.getName().equals(this.getName())
                     &&c.cardNumber==this.cardNumber
                     &&c.PIN==this.PIN
                     )
                 return true;
          }
          return false;
       }
    public String toString()
    {
```

```
            return
"CallingCard[name="+getName()+",cardNumber="+cardNumber+",PIN="+PIN+"]";
        }
    }
```

4.

```
    /**
     * DriverLicense
     * @author ********
     *
     */
    public class DriverLicense extends CallingCard {
    private int expirationYear;
        public DriverLicense(String name, String num, String p ,int year)
        {
            super(name,num,p);
            expirationYear=year;
        }
        public String toString()
        {
            return
"DriverLicense[name="+getName()+",cardNumber="+getCardNumber()
                  +",PIN="+getPIN()+",year="+expirationYear+"]";
        }
        public String format()
        {
            return super.format()+" year:"+expirationYear;
        }
         public boolean equals(Object o)
          {
              if(this==o)
                  return true;
              if(o==null)
                  return false;
              if(o.getClass().getName()==this.getClass().getName())
              {
                  DriverLicense c=(DriverLicense)o;
                  if(c.getName().equals(this.getName())
                          &&c.getCardNumber()==this.getCardNumber()
                          &&c.getPIN()==this.getPIN()
                          &&c.expirationYear==this.expirationYear
                          )
                      return true;
              }
              return false;
          }
        }
```

5.

```
/**
 *测试类
 * @author ********
 *
 */
import org.junit.Test;
public class testJUnit {
    @Test
    public void testCard() {
        Card ic1=new Card("peter");
        System.out.println("ic1"+ic1.toString());
        System.out.println("ic1"+ic1.format());
        Card ic2=new Card("peter");
        System.out.println("ic2"+ic2.toString());
        System.out.println("ic2"+ic2.format());
        System.err.println("ic1 equals ic2 "+ic1.equals(ic2));
    }
    @Test
    public void testIDCard() {
        IDCard ic1=new IDCard("peter", "123");
        System.out.println("ic1"+ic1.toString());
        System.out.println("ic1"+ic1.format());
        IDCard ic2=new IDCard("peter", "123");
        System.out.println("ic2"+ic2.toString());
        System.out.println("ic2"+ic2.format());
        System.err.println("ic1 equals ic2 "+ic1.equals(ic2));
    }
    @Test
    public void testCallingCard() {
        CallingCard ic1=new CallingCard("peter", "123","000");
        System.out.println("ic1 "+ic1.toString());
        System.out.println(ic1.format());
        CallingCard ic2=new CallingCard("peter", "123","001");
        System.out.println("ic2 "+ic2.toString());
        System.err.println("ic1 equals ic2 "+ic1.equals(ic2));
    }
    @Test
    public void testDriverLicense() {
        DriverLicense ic1=new DriverLicense("peter", "123","000",3);
        System.out.println("ic1 "+ic1.toString());
        System.out.println(ic1.format());
        DriverLicense ic2=new DriverLicense("peter", "123","000",3);
        System.out.println("ic2 "+ic2.toString());
        System.err.println("ic1 equals ic2 "+ic1.equals(ic2));
    }
}
```

7.5 Experimental result（实验结果）

实验结果如图 7-6～图 7-9 所示。

1） testCard

```
ic1Card[name=peter]
ic1Card holder: peter
ic2Card[name=peter]
ic2Card holder: peter
ic1 equals ic2 true
```

Fig7-6. testCard

2） testIDCard

```
ic1IDCard[name=peter,IDNumber=123]
ic1Card holder: peter IDNumber:123
ic2IDCard[name=peter,IDNumber=123]
ic2Card holder: peter IDNumber:123
ic1 equals ic2 true
```

Fig7-7. testIDCard

3） testCallingCard

```
ic1 CallingCard[name=peter,cardNumber=123,PIN=000]
Card holder: peter cardNumber:123 PIN:000
ic2 CallingCard[name=peter,cardNumber=123,PIN=001]
ic1 equals ic2 false
```

Fig7-8. testCallingCard

4） testDriverLicense

```
ic1 DriverLicense[name=peter,cardNumber=123,PIN=000,year=3]
Card holder: peter cardNumber:123 PIN:000 year:3
ic2 DriverLicense[name=peter,cardNumber=123,PIN=000,year=3]
ic1 equals ic2 true
```

Fig7-9. testDriverLicense

第 8 章　Graphical User Interfaces（图形用户接口）

8.1　Key points of this chapter（本章要点）

1. Frame Windows（框架窗口）

A graphical application shows information inside a frame: a window with a title bar.

用一个图形化应用程序显示一个框架内的信息：带有标题栏的窗口。

1）Displaying a Frame（显示框架）

```
JFrame frame = new JFrame();
final int FRAME_WIDTH = 300;
final int FRAME_HEIGHT = 400;
frame.setSize(FRAME_WIDTH, FRAME_HEIGHT);
frame.setTitle("An empty frame");
frame.setDefaultCloseOperation(JFrame.EXIT_ON_CLOSE);
frame.setVisible(true);
```

2）Adding User-Interface Components to a Frame（在框架中添加用户界面组件）

An empty frame is not very interesting. You will want to add some user-interface components, such as buttons and text labels.

一个空的框架没有意义，你可能想要给它加一些用户界面组件，像按钮、文本标签。

```
JPanel panel = new JPanel();
panel.add(button);
panel.add(label);
frame.add(panel);
```

2. Events and Event Handling（事件和事件处理）

In a program with a modern graphical user interface, the user is in control. The user can use both the mouse and the keyboard and can manipulate many parts of the user interface in any desired order. The program must react to the user commands in whatever order they arrive.

现代的图形用户界面程序都是由用户来操作控制的，用户界面的大部分功能都能由用户通过鼠标和键盘来操作。

1） Listening to Events（事件侦听）

Every program must indicate which events it needs to receive. It does that by installing event listener objects. These objects are instances of classes that you must provide. The methods of your event listener classes contain the instructions that you want to have executed when the events occur. For example, Button listeners must belong to a class that implements the ActionListener interface:

每一个程序都通过安装事件侦听器对象来表示出它能接受哪些事件，这些对象是必须提供的类的实例，事件监听器的方法包含当事件发生时要执行的指令。比如 Button 侦听器必须属于实现了 ActionListener 接口的类：

```
public interface ActionListener
{
void actionPerformed(ActionEvent event);
}
```

Once the listener class has been declared, we need to construct an object of the class and add it to the button:

一旦侦听器类已被声明，我们需要构造一个类的对象并将其添加到按钮上：

```
ActionListener listener = new ClickListener();
button.addActionListener(listener);
```

2） Using Inner Classes for Listeners（使用内部类作为侦听器）

It is common to implement listener classes as inner classes like this:

下面是一个常见的通过内部类实现侦听器的例子：

```
public class ButtonFrame2 extends JFrame
{
    . . .
    // This inner class is declared inside the frame class
    class ClickListener implements ActionListener
    {
    . . .
    }
    private void createComponents()
    {
      button = new JButton("Click me!");
      ActionListener listener = new ClickListener();
      button.addActionListener(listener);
      . . .
    }
}
```

3. Swing User-Interface Components（Swing 用户界面组件）

Swing 用户界面组件层次结构的一部分如图 8-1 所示。

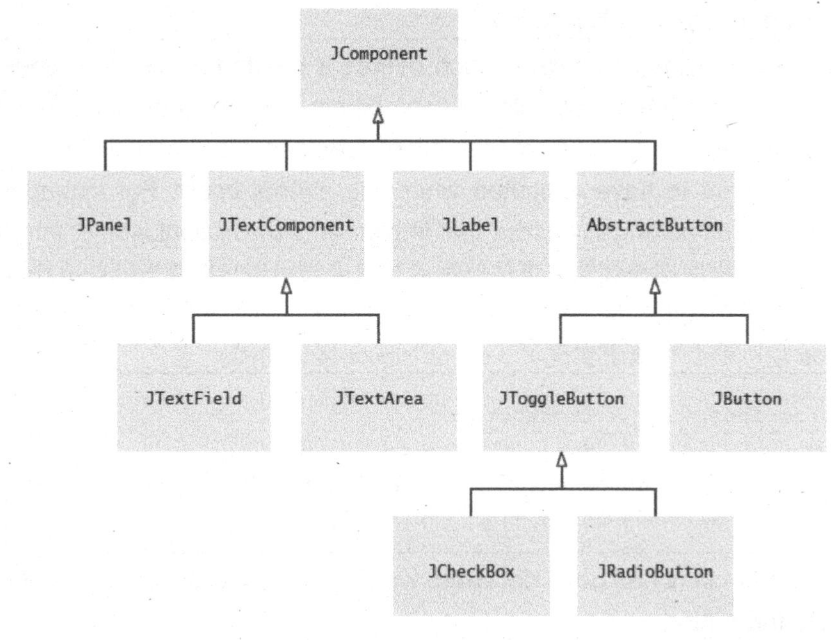

Fig8-1. A Part of the Hierarchy of Swing User-Interface Components

4. Creating Drawings（实现画图）

You cannot draw directly onto a frame. Instead, you add a component to the frame and draw on the component. To do so, extend the JComponent class and override its paintComponent method.

你不能直接在框架上画图，而是将组件添加到框架内并在组件上画图。为此，请扩展 JComponent 类并重写其 paintComponent 方法。

```
public class ChartComponent extends JComponent
{
   public void paintComponent(Graphics g)
   {
     Drawing instructions
   }
}
```

5. Layout Management（管理布局）

A flow layout simply arranges its components from left to right and starts a new row when there is no more room in the current row.
Another commonly used layout manager is the border layout. The border layout groups components into five areas: center, north, south, west, and east. Each area can hold a single component, or it can be empty.
The grid layout manager arranges components in a grid with a fixed number of rows

and columns. All components are resized so that they all have the same width and height.
流式布局简单地从左到右排列其组件，并在当前行中没有更多空间时启动新行。 另一个常用布局是边界布局，边界布局将组成部分分为五个区域：中心，北，南，西和东。每个区域可以容纳单个组件，也可以为空。 网格布局将组件布置在具有固定数目的行和列的网格中。自动调整所有组件的大小，使它们都具有相同的宽度和高度。

6．Laying Out a User Interface（用户界面布局）

Step 1 Make a sketch of your desired component layout; Step 2 Find groupings of adjacent components with the same layout; Step 3 Identify layouts for each group; Step 4 Group the groups together; Step 5 Write the code to generate the layout.
第一步：制作所需组件布局的草图； 第二步：查找同一布局的相邻组件的分组； 第三步：确定每个分组的布局； 第四步：把各个分组组合在一起； 第五步：编写生成布局的代码。

7．Menus（菜单）

1）	menu bar（菜单栏）
	At the top of the frame is a menu bar that contains the top-level menus. Each menu is a collection of menu items and submenus.
	在框架的顶部是一个包含顶层菜单的菜单栏。每个菜单由一组菜单项和子菜单组成。

```
JMenuBar menuBar = new JMenuBar();
setJMenuBar(menuBar);
```

2）	Menus（菜单）
	Menus are then added to the menu bar:
	添加菜单到菜单栏中：

```
JMenu fileMenu = new JMenu("File");
JMenu fontMenu = new JMenu("Font");
menuBar.add(fileMenu);
menuBar.add(fontMenu);
```

3）	menu items（菜单项）
	You add menu items and submenus with the add method:
	通过 add 方法实现添加菜单项和子菜单到菜单中。

```
JMenuItem exitItem = new JMenuItem("Exit");
fileMenu.add(exitItem);
JMenu styleMenu = new JMenu("Style");
fontMenu.add(styleMenu); // A submenu
```

4）	a listener to each menu item（每个菜单项的侦听器）
	When the user selects a menu item, the menu item sends an action event. Therefore, you want to add a listener to each menu item: 当用户选择一个菜单项时，菜单项递交一个动作事件。因此，你需要在每个菜单项中添加一个侦听器。

```
ActionListener listener = new ExitItemListener();
exitItem.addActionListener(listener);
```

8. Using Timer Events for Animations（使用动画的定时器事件）

The Timer class in the javax.swing package generates a sequence of action events, spaced at even time intervals. For example:

在 javax.swing 包的定时器类生成一个序列的动作事件，时间间隔要均匀。例如：

```
class MyListener implements ActionListener
{
   public void actionPerformed(ActionEvent event)
   {
   Action that is executed at each timer event
   }
}
MyListener listener = new MyListener();
Timer t = new Timer(interval, listener);
t.start();
```

9. Mouse Events（鼠标事件）

A mouse listener must implement the MouseListener interface, which contains the following five methods:

鼠标监听器必须实现 MouseListener 接口，它包含以下五种方法：

```
public interface MouseListener
{
   void mousePressed(MouseEvent event);
   // Called when a mouse button has been pressed on a component
   void mouseReleased(MouseEvent event);
   // Called when a mouse button has been released on a component
   void mouseClicked(MouseEvent event);
   // Called when the mouse has been clicked on a component
   void mouseEntered(MouseEvent event);
   // Called when the mouse enters a component
   void mouseExited(MouseEvent event);
   // Called when the mouse exits a component
```

```
    }
```

8.2　Example（例题）

1.	We will build a practical application with a graphical user interface. A frame displays the amount of money in a bank account. Whenever the user clicks a button, 5 percent interest is added, and the new balance is displayed. 我们将使用图形用户界面构建实际应用程序。框架显示银行账户中的金额。每当用户单击按钮时，将添加 5%的利息，并显示新的余额。

文本字段的应用程序如图 8-2 所示。

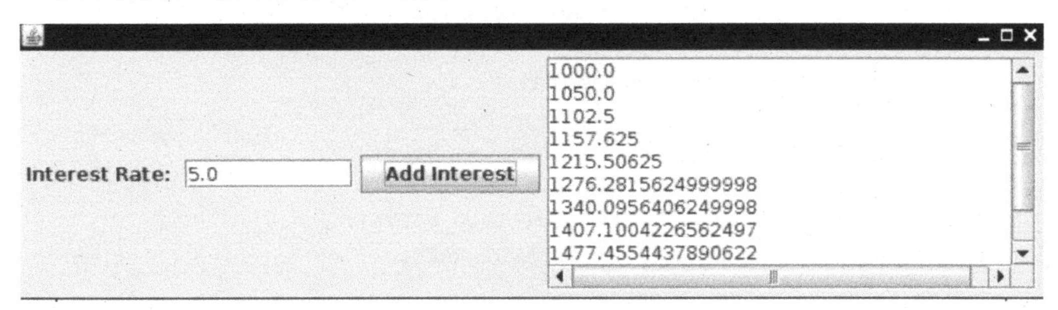

Fig8-2.　An Application with a Text Field

1）	InvestmentFrame.java

```
import java.awt.event.ActionEvent;
import java.awt.event.ActionListener;
import javax.swing.JButton;
import javax.swing.JFrame;
import javax.swing.JLabel;
import javax.swing.JPanel;
import javax.swing.JScrollPane;
import javax.swing.JTextArea;
import javax.swing.JTextField;
/**
   A frame that shows the growth of an investment with variable interest,
   using a text area.
*/
public class InvestmentFrame extends JFrame
{
   private static final int FRAME_WIDTH = 400;
   private static final int FRAME_HEIGHT = 250;
   private static final int AREA_ROWS = 10;
   private static final int AREA_COLUMNS = 30;
   private static final double DEFAULT_RATE = 5;
   private static final double INITIAL_BALANCE = 1000;
   private JLabel rateLabel;
```

```
    private JTextField rateField;
    private JButton button;
    private JTextArea resultArea;
    private double balance;
    public InvestmentFrame()
    {
      balance = INITIAL_BALANCE;
      resultArea = new JTextArea(AREA_ROWS, AREA_COLUMNS);
      resultArea.setText(balance + "\n");
      resultArea.setEditable(false);
      createTextField();
      createButton();
      createPanel();
      setSize(FRAME_WIDTH, FRAME_HEIGHT);
    }
    private void createTextField()
    {
      rateLabel = new JLabel("Interest Rate: ");
      final int FIELD_WIDTH = 10;
      rateField = new JTextField(FIELD_WIDTH);
      rateField.setText("" + DEFAULT_RATE);
    }
    class AddInterestListener implements ActionListener
    {
      public void actionPerformed(ActionEvent event)
      {
        double rate = Double.parseDouble(rateField.getText());
        double interest = balance * rate / 100;
        balance = balance + interest;
        resultArea.append(balance + "\n");
      }
    }
    private void createButton()
    {
      button = new JButton("Add Interest");

      ActionListener listener = new AddInterestListener();
      button.addActionListener(listener);
    }
    private void createPanel()
    {
      JPanel panel = new JPanel();
      panel.add(rateLabel);
      panel.add(rateField);
      panel.add(button);
      JScrollPane scrollPane = new JScrollPane(resultArea);
      panel.add(scrollPane);
```

```
        add(panel);
    }
}
```

2） InvestmentViewer.java

```
import javax.swing.JFrame;
/**
  This program displays the growth of an investment.
*/
public class InvestmentViewer
{
  public static void main(String[] args)
  {
    JFrame frame = new InvestmentFrame();
    frame.setDefaultCloseOperation(JFrame.EXIT_ON_CLOSE);
    frame.setVisible(true);
  }
}
```

2. This frame has a menu with commands to change the font of a text sample.
此框架包含用于更改文本字体的命令的菜单。

菜单如图 8-3～图 8-6 所示。

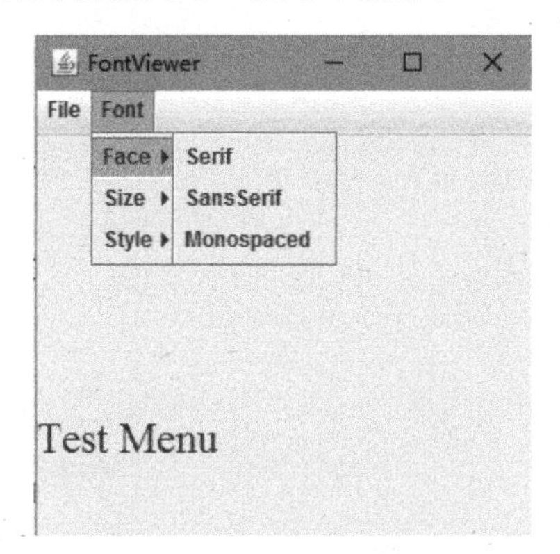

Fig8-3. Pull-Down Menus 1

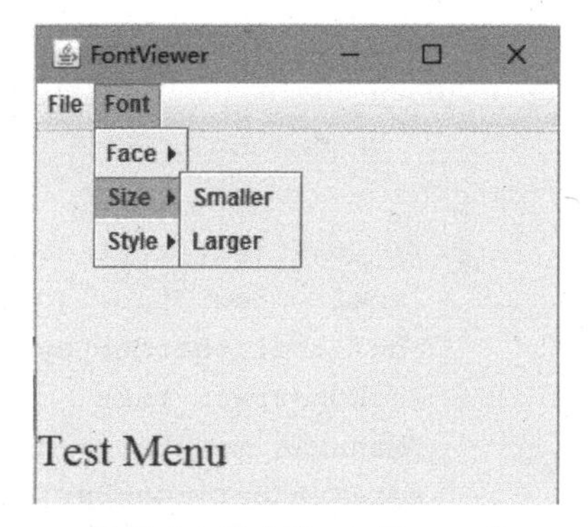

Fig8-4. Pull-Down Menus 2

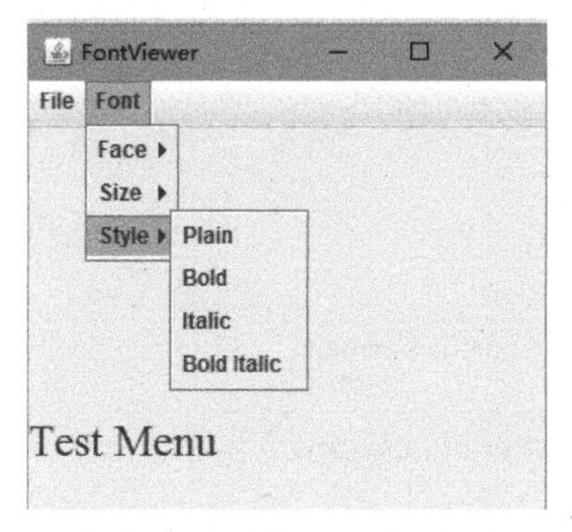

Fig8-5. Pull-Down Menus 3

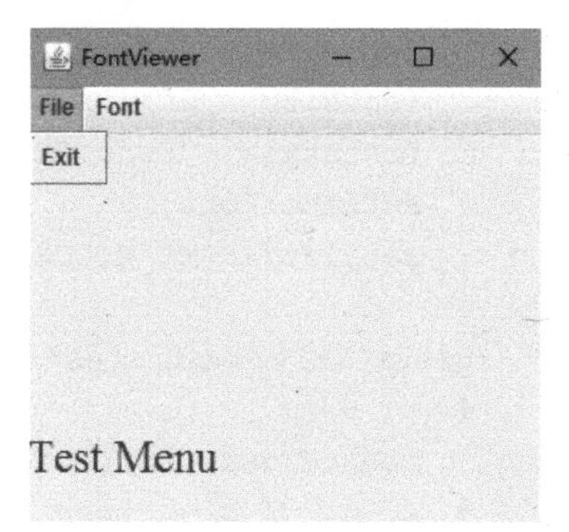

Fig8-6. Pull-Down Menus 4

1）FontFrame.java

```
import java.awt.BorderLayout;
import java.awt.Font;
import java.awt.event.ActionEvent;
import java.awt.event.ActionListener;
import javax.swing.JFrame;
import javax.swing.JLabel;
import javax.swing.JMenu;
import javax.swing.JMenuBar;
import javax.swing.JMenuItem;
/**
  This frame has a menu with commands to change the font
  of a text sample.
*/
public class FontFrame extends JFrame
{
  private static final int FRAME_WIDTH = 300;
  private static final int FRAME_HEIGHT = 400;
  private JLabel label;
  private String facename;
  private int fontstyle;
  private int fontsize;
  /**
    Constructs the frame.
  */
  public FontFrame ()
  {
    // Construct text sample
    label = new JLabel("Test Menu");
    add(label, BorderLayout.CENTER);
    // Construct menu
    JMenuBar menuBar = new JMenuBar();
    setJMenuBar(menuBar);
    menuBar.add(createFileMenu());
    menuBar.add(createFontMenu());
    facename = "Serif";
    fontsize = 24;
    fontstyle = Font.PLAIN;
    setLabelFont();
    setSize(FRAME_WIDTH, FRAME_HEIGHT);
  }
  class ExitItemListener implements ActionListener
  {
    public void actionPerformed(ActionEvent event)
    {
      System.exit(0);
```

```
      }
   }
   /**
      Creates the File menu.
      @return the menu
   */
   public JMenu createFileMenu()
   {
      JMenu menu = new JMenu("File");
      JMenuItem exitItem = new JMenuItem("Exit");
      ActionListener listener = new ExitItemListener();
      exitItem.addActionListener(listener);
      menu.add(exitItem);
      return menu;
   }
   /**
      Creates the Font submenu.
      @return the menu
   */
   public JMenu createFontMenu()
   {
      JMenu menu = new JMenu("Font");
      menu.add(createFaceMenu());
      menu.add(createSizeMenu());
      menu.add(createStyleMenu());
      return menu;
   }
   /**
      Creates the Face submenu.
      @return the menu
   */
   public JMenu createFaceMenu()
   {
      JMenu menu = new JMenu("Face");
      menu.add(createFaceItem("Serif"));
      menu.add(createFaceItem("SansSerif"));
      menu.add(createFaceItem("Monospaced"));
      return menu;
   }
   /**
      Creates the Size submenu.
      @return the menu
   */
   public JMenu createSizeMenu()
   {
      JMenu menu = new JMenu("Size");
      menu.add(createSizeItem("Smaller", -1));
```

```
      menu.add(createSizeItem("Larger", 1));
      return menu;
   }
   /**
      Creates the Style submenu.
      @return the menu
   */
   public JMenu createStyleMenu()
   {
      JMenu menu = new JMenu("Style");
      menu.add(createStyleItem("Plain", Font.PLAIN));
      menu.add(createStyleItem("Bold", Font.BOLD));
      menu.add(createStyleItem("Italic", Font.ITALIC));
      menu.add(createStyleItem("Bold Italic", Font.BOLD
           + Font.ITALIC));
      return menu;
   }
   /**
      Creates a menu item to change the font face and set its action listener.
      @param name the name of the font face
      @return the menu item
   */
   public JMenuItem createFaceItem(final String name)
   {
      class FaceItemListener implements ActionListener
      {
         public void actionPerformed(ActionEvent event)
         {
            facename = name;
            setLabelFont();
         }
      }
      JMenuItem item = new JMenuItem(name);
      ActionListener listener = new FaceItemListener();
      item.addActionListener(listener);
      return item;
   }
   /**
      Creates a menu item to change the font size
      and set its action listener.
      @param name the name of the menu item
      @param increment the amount by which to change the size
      @return the menu item
   */
   public JMenuItem createSizeItem(String name, final int increment)
   {
      class SizeItemListener implements ActionListener
```

```
            {
               public void actionPerformed(ActionEvent event)
               {
                  fontsize = fontsize + increment;
                  setLabelFont();
               }
            }
            JMenuItem item = new JMenuItem(name);
            ActionListener listener = new SizeItemListener();
            item.addActionListener(listener);
            return item;
         }
         /**
            Creates a menu item to change the font style
            and set its action listener.
            @param name the name of the menu item
            @param style the new font style
            @return the menu item
         */
         public JMenuItem createStyleItem(String name, final int style)
         {
            class StyleItemListener implements ActionListener
            {
               public void actionPerformed(ActionEvent event)
               {
                  fontstyle = style;
                  setLabelFont();
               }
            }
            JMenuItem item = new JMenuItem(name);
            ActionListener listener = new StyleItemListener();
            item.addActionListener(listener);
            return item;
         }
         /**
            Sets the font of the text sample.
         */
         public void setLabelFont()
         {
            Font f = new Font(facename, fontstyle, fontsize);
            label.setFont(f);
         }
      }
```

2） FontViewer.java

```
      import javax.swing.JFrame;
```

```
/**
   This program uses a menu to display font effects.
*/
public class FontViewer
{
   public static void main(String[] args)
   {
      JFrame frame = new FontFrame ();
      frame.setDefaultCloseOperation(JFrame.EXIT_ON_CLOSE);
      frame.setTitle("FontViewer");
      frame.setVisible(true);
   }
}
```

8.3 Experimental contents（实验内容）

<table>
<tr><td rowspan="2">1.</td><td>1) Write a main program that displays a single frame with the title “My First Frame”. Set the size to 800 by 800. Make the frame visible.
2) Create a panel by using the JPanel constructor and add it to the frame. Use Color.RED (a constant in the java.awt package) along with the setBackground() method in JPanel, to set the color of the panel. Add a JButton and a JLabel to the panel before adding the panel to the frame. Display the results.
3) The goal of this problem is to repackage most of the previous code as a class that extends JFrame. Build a class called MyCustomFrame that extends JFrame. The class should contain instance variables for the button and the label, as well as constants for the frame width and height. The constructor should call a private helper method, createComponents(), that instantiates the button, the label, and the panel. The helper method should also add the button and the label to the panel, and add the panel to the frame. After creating the components, the constructor should set the size of the frame.
4) Build a separate class called ClickListener that implements the ActionListener interface. Add an actionPerformed() method that prints the message “Button was clicked.” using System.out.println(). Create a ClickListener object and register the object with the button by invoking addActionListener().</td></tr>
<tr><td>1）编写一个主程序，显示标题为“My First Frame”的单个框架。将大小设置为 800*800。设置框架可见。
2）使用 JPanel 构造函数创建一个面板，并将其添加到框架里。使用 Color.RED（java.awt 包中的常量）以及 JPanel 中的 setBackground()方法来设置面板的颜色。在将面板添加到框架之前，向面板添加 JButton 和 JLabel，并显示结果。
3）本题的目的是将以前的大多数代码重新打包为一个扩展 JFrame 的类。构建一个名为 MyCustomFrame 的类来扩展 JFrame。该类应包含按钮和标签的实例变量，</td></tr>
</table>

以及框架宽度和高度的常量。构造函数应该调用一个私有方法 CreateComponents()，它能实例化按钮，标签和面板。该方法还应将按钮和标签添加到面板，并将面板添加到框架。创建组件后，构造函数应设置框架的大小。

4）构建一个名为 ClickListener 的类，实现 ActionListener 接口。添加一个 actionPerformed()方法，它使用 System.out.println()输出消息“Button was clicked”。创建一个 ClickListener 对象，并通过调用 addActionListener()向按钮注册对象。

Test the program by clicking the button. When the program is working correctly, convert the ClickListener class to an inner class in MyCustomFrame and test the program again.

通过单击按钮测试程序。当程序正常工作时，将 ClickListener 类转换为 MyCustomFrame 中的内部类并再次测试程序。

5) Add a JTextField and a JTextArea to the frame. Initialize the field to an empty string and the text area to “||”. Modify the program so that the contents of the field are appended to the area each time the user clicks the button. For example, if the user enters “xxx” in the field and clicks the button three times, the area contains “||xxx||xxx||xxx||”.

5）将 JTextField 和 JTextArea 添加到框架。将字段初始化为空字符串，将文本区域初始化为“||”。修改程序，以便每次用户单击按钮时，将字段的内容附加到该区域。例如，如果用户在字段中输入“xxx”并单击按钮三次，则该区域包含“|| xxx || xxx || xxx ||”。

Test your class with the viewer code below.

使用下面的代码测试你的类。

FontFrame 组件如图 8-7 所示。

Fig8-7. The Components of the FontFrame

```
import javax.swing.JFrame;
public class MyCustomFrameViewer
{
   public static void main(String[] args)
   {
      MyCustomFrame frame = new MyCustomFrame();
      frame.setTitle("My first frame");
```

```
        frame.setDefaultCloseOperation(JFrame.EXIT_ON_CLOSE);
        frame.setVisible(true);
    }
}
```

<table>
<tr><td rowspan="2">2.</td><td>Write an application with a Color menu and menu items labeled “Red”, “Green”, and “Blue” that change the background color of a panel in the center of the frame to red, green, or blue.</td></tr>
<tr><td>编写一个应用程序，它有一个设置颜色的菜单和内容为“Red”，“Green”和“Blue”的三个菜单项，通过选中菜单项将框架中心面板的背景颜色改为红、绿、蓝。</td></tr>
</table>

8.4 Experimental steps（实验步骤）

1.

```
import java.awt.Color;
import java.awt.Dimension;
import java.awt.event.ActionEvent;
import java.awt.event.ActionListener;
import javax.swing.JButton;
import javax.swing.JFrame;
import javax.swing.JLabel;
import javax.swing.JPanel;
import javax.swing.JTextArea;
import javax.swing.JTextField;
public class MyCustomFrame extends JFrame {
    private JButton button;
    private JLabel label;
    private JPanel panel;
    private JTextField text;
    private JTextArea area;
    public MyCustomFrame(){
        createComponents();
    }
    public void createComponents(){
        button=new JButton("button");
        ActionListener listener=new ClickListener();
        button.addActionListener(listener);
        label=new JLabel("label");
        panel=new JPanel();
        text=new JTextField("");
        text.setPreferredSize(new Dimension(300,30));
        area=new JTextArea("||");
```

```
        area.setPreferredSize(new Dimension(300,300));
        panel.add(button);
        panel.add(label);
        panel.add(text);
        panel.add(area);
        panel.setBackground(Color.RED);
        add(panel);
    }
    public class ClickListener implements ActionListener{
        @Override
        public void actionPerformed(ActionEvent arg0) {
            // TODO Auto-generated method stub
            System.out.println("Button was clicked.");
            String str=text.getText();
            area.append(str+"||");
        }
    }
}
```

2.

```
import javax.swing.JFrame;
public class MyCustomFrameViewer {
    public static void main(String[] args) {
        // TODO Auto-generated method stub
        MyCustomFrame frame = new MyCustomFrame();
        frame.setTitle("My first frame");
        frame.setSize(800, 800);
    frame.setDefaultCloseOperation(JFrame.EXIT_ON_CLOSE);
        frame.setVisible(true);
    }
}
```

3.

```
import java.awt.Color;
import java.awt.Frame;
import java.awt.event.ActionEvent;
import java.awt.event.ActionListener;
import javax.swing.JFrame;
import javax.swing.JMenu;
import javax.swing.JMenuBar;
import javax.swing.JMenuItem;
import javax.swing.JPanel;
/**
 * 颜色菜单
 * @author *****
 *
 */
```

```
@SuppressWarnings("serial")
public class MenuOfColor extends JFrame {
    private JPanel panel;
    public MenuOfColor(){
        JMenu colorMenu=new JMenu("color");
        ActionListener listener=new ClickListener();
        JMenuItem red=new JMenuItem("red");
        red.addActionListener(listener);
        colorMenu.add(red);
        JMenuItem green=new JMenuItem("green");
        green.addActionListener(listener);
        colorMenu.add(green);
        JMenuItem blue=new JMenuItem("blue");
        blue.addActionListener(listener);
        colorMenu.add(blue);
        JMenuBar bar=new JMenuBar();
        bar.add(colorMenu);
        setJMenuBar(bar);
        panel=new JPanel();
        add(panel);
    }
    class ClickListener implements ActionListener{
        @Override
        public void actionPerformed(ActionEvent arg0) {
            // TODO Auto-generated method stub
            String str=arg0.getActionCommand();
            if(str.equals("red")){
                panel.setBackground(Color.RED);
            }
            else if(str.equals("green")){
                panel.setBackground(Color.GREEN);
            }
            else if(str.equals("blue")){
                panel.setBackground(Color.BLUE);
            }
        }
    }
    public static void main(String[] args) {
        // TODO Auto-generated method stub
        MenuOfColor m1=new MenuOfColor();
        m1.setTitle("color menu");
        m1.setSize(300,200);
        m1.setDefaultCloseOperation(JFrame.EXIT_ON_CLOSE);
        m1.setVisible(true);
    }
}
```

8.5 Experimental result（实验结果）

实验结果如图 8-8～图 8-11 所示。

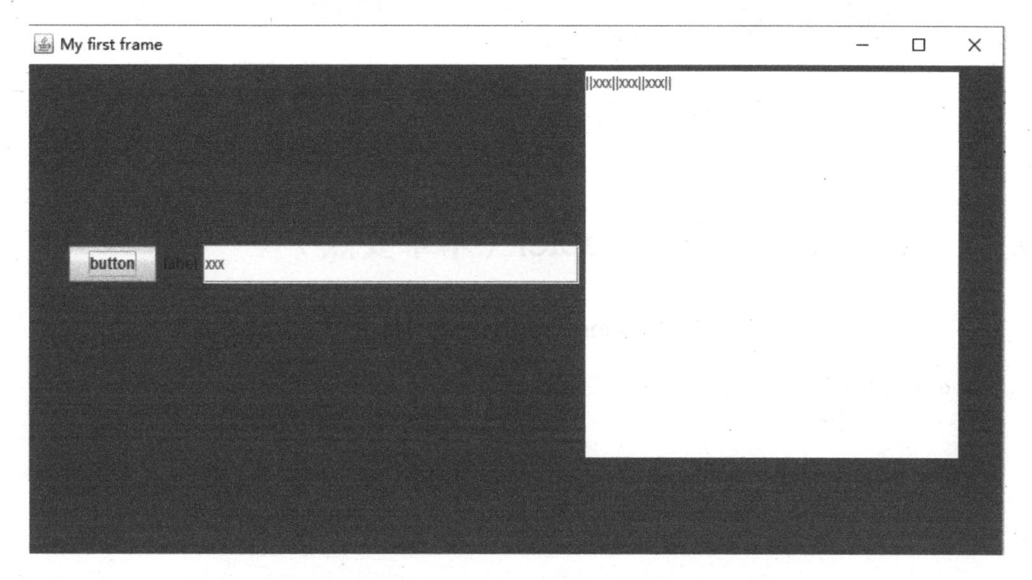

Fig8-8.　The result of The Components of the FontFrame

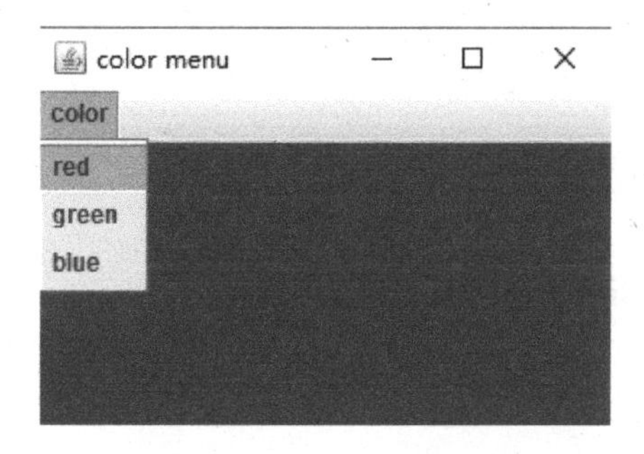

Fig8-9.　The result of Menu 1

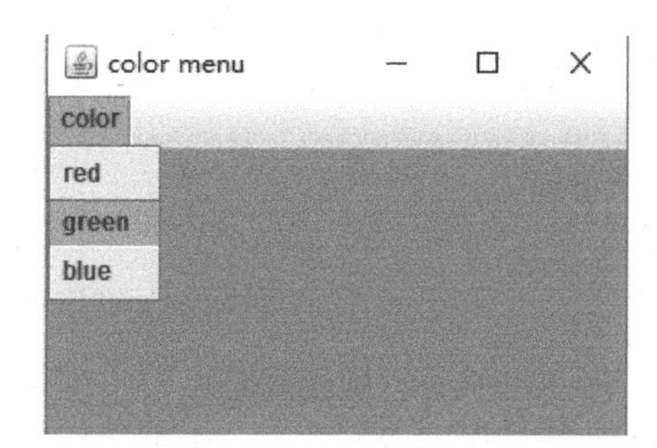

Fig8-10.　The result of Menu 2

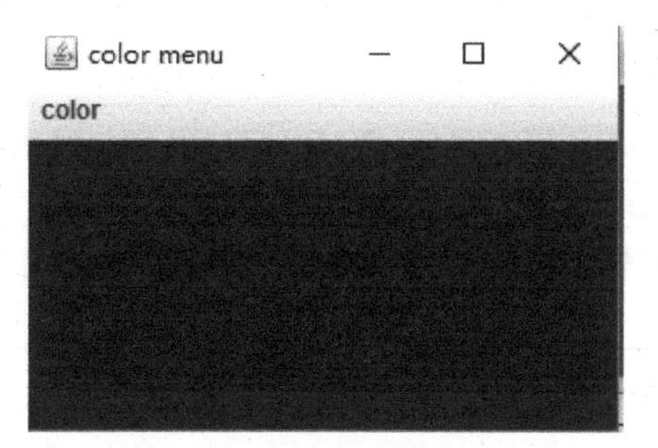

Fig8-11.　The result of Menu 3

第 9 章　the Java Collections Framework（Java 集合框架）

9.1　Key points of this chapter（本章要点）

1．An Overview of the Collections Framework（集合框架视图）

Java 集合框架的接口和类如图 9-1 所示。

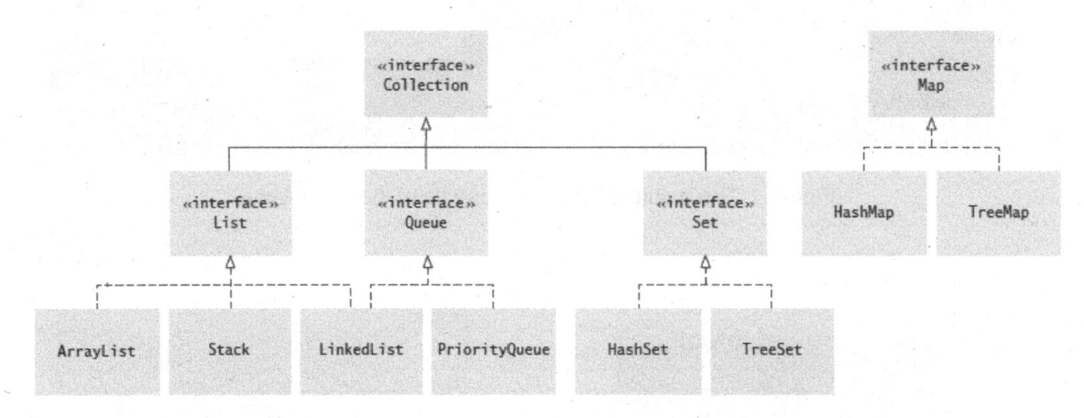

Fig9-1.　Interfaces and Classes in the Java Collections Framework

2．ArrayList（数组列表）

Stores a list of items in a dynamically sized array.
把一列数据项存储在动态数组中。

3．LinkedLists（链接列表）

A linked list uses a sequence of nodes. A node is an object that stores an element and references to the neighboring nodes in the sequence.
链接列表使用节点序列。每个节点都是一个对象，其中存储了序列中一个元素和对邻接节点的引用。

链接列表如图 9-2 所示。

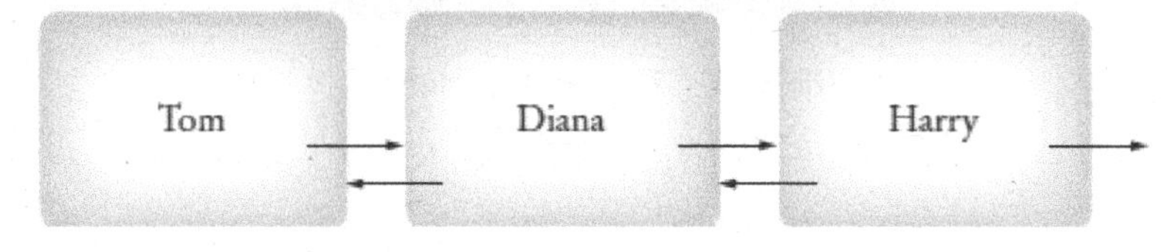

Fig9-2.　A Linked List

1）The LinkedList Class of the Java Collections Framework（Java 集合框架中的 LinkedList）

```
LinkedList<String> list = new LinkedList<String>();
list.addLast("Harry");
list.addFirst("Sally");
list.getFirst();
list.getLast();
```

2）List Iterators（列表迭代器）

An iterator encapsulates a position anywhere inside the linked list.

一个迭代器封装链表中每个元素的位置。

```
LinkedList<String> employeeNames = . . .;
ListIterator<String> iterator = employeeNames.listIterator();
```

You traverse all elements in a linked list of strings with the following loop:

使用下面的循环，你可以遍历链表的字符串中所有的元素：

```
while (iterator.hasNext())
{
    String name = iterator.next();
    Do something with name
}
```

As a shorthand, if your loop simply visits all elements of the linked list, you can use the “for each” loop:

作为一个快速方式，如果循环简单地访问链表的所有元素，你可以使用“for each”循环：

```
for (String name : employeeNames)
{
Do something with name
}
```

4. Sets（集合）

There is an essential difference between arbitrary collections and sets. A set does not admit duplicates. If you add an element to a set that is already present, the insertion is ignored.

在其他集合与 set 集合之间有一个重要的不同点。一个 set 集合不允许有相同的元素。如果向 set 集合中添加一个已有数据，这个新添加的元素不会出现在集合中。

1）HashSet：Stores data in a Hash Table. Set elements are grouped into smaller collections of elements that share the same characteristic.

Hash 集合：存储数据在 Hash 表。集合元素被分为较小的元素集合，这些元素

共享相同的特征。

设置接口如图 9-3 所示。

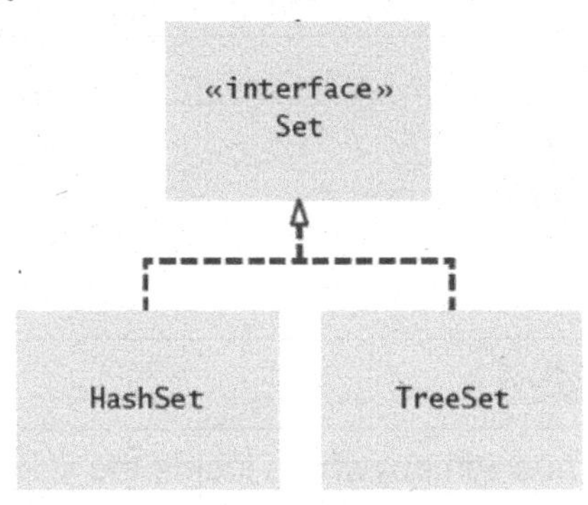

Fig9-3. Set Interface

Hash 表如图 9-4 所示。

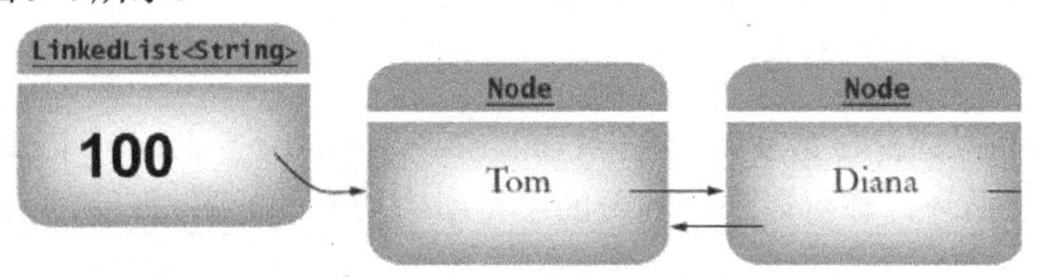

Fig9-4. Hash Table

int hashCode()：Returns the hash code value for this collection. While the Collection interface adds no stipulations to the general contract for the Object.hashCode method, programmers should take note that any class that overrides the Object.equals method must also override the Object.hashCode method in order to satisfy the general contract for the Object.hashCode method. In particular, c1.equals(c2) implies that c1.hashCode()==c2.hashCode().
int hashCode()：返回这个集合的哈希码值。当 collection 接口没有为 object.hashCode()方法的常规协议添加任何约束时，程序员应该注意，为了满足 object.hashcode()方法的一般约定，任何类，覆盖 Object.equals()方法也必须重写为 object.hashcode()方法。需要特别指出的是，c1.equals(c2)暗示着 c1.hashCode()==c2.hashCode().

```
Set<String> names = new HashSet<String>();
names.add("Romeo");
names.remove("Juliet");
if (names.contains("Juliet")) . . .
Iterator<String> iter = names.iterator();
while (iter.hasNext())
{
   String name = iter.next();
   Do something with name
}
```

2）	TreeSet: Nodes are not arranged in a linear sequence but in a tree shape, Stores data in a Binary Tree.

TreeSet：节点是分布在一个树状结构而不是一个线性序列结构上。使用 TreeSet 存储数据，数据存储在一个二叉树上，如图 9-5 所示。

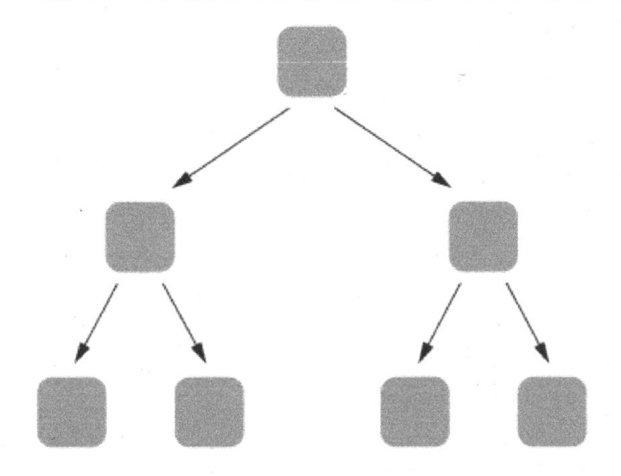

Fig9-5.　TreeSet

```
Set<String> names = new TreeSet<String>();
names.add("Romeo");
names.remove("Juliet");
if (names.contains("Juliet")) . . .
Iterator<String> iter = names.iterator();
while (iter.hasNext())
{
   String name = iter.next();
   Do something with name
}
```

5. Maps

A map allows you to associate elements from a key set with elements from a value collection. The HashMap and TreeMap classes both implement the Map interface.

一个 map 存储值是以键值对 key-value 的结构，hashmap 和 treemap 都实现 map 接口，如图 9-6 所示。

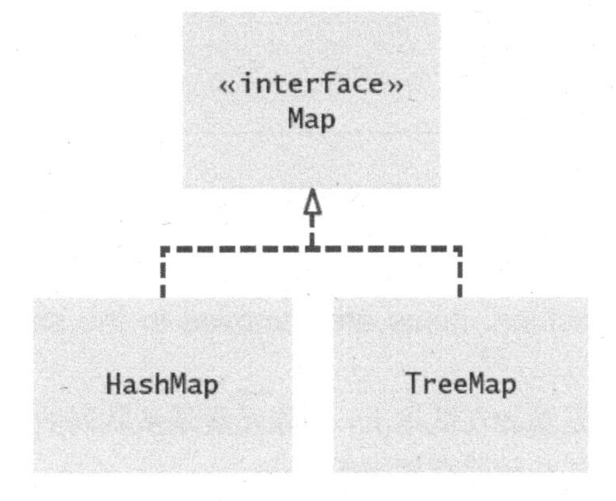

Fig9-6.　Map interface

应用 Map 如图 9-7 所示。

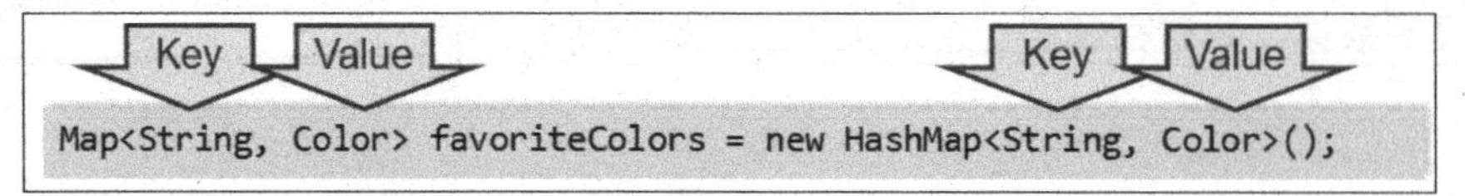

Fig9-7. Using Map

```
Map<String, Integer> scores;
scores = new TreeMap<String, Integer>();
scores.put("Harry", 90);
scores.put("Sally", 95);
int n = scores.get("Sally");
Integer n2 = scores.get("Diana");
for (String key : scores.keySet())
{
    Integer value = scores.get(key);
    . . .
}
scores.remove("Sally");
```

6. Stacks, Queues, and Priority Queues（栈，队列和优先队列）

1）	Stacks（栈）
	A stack lets you insert and remove elements only at one end, traditionally called the top of the stack. New items can be added to the top of the stack. Items are removed from the top of the stack as well. Therefore, they are removed in the order that is opposite from the order in which they have been added, called last-in, first-out or LIFO order.
	一个栈可以在末尾插入和移除一个元素，俗称栈顶元素。在栈顶添加一个新的元素，同时也是删除栈顶元素。因此，栈的添加和删除位置是倒置的，后进先出。

```
Stack<String> s = new Stack<String>();
s.push("A"); s.push("B"); s.push("C");
while (s.size() > 0)
{
System.out.print(s.pop() + " "); // Prints C B A
}
```

2）	Queues（队列）
	A queue lets you add items to one end of the queue (the tail) and remove them from the other end of the queue (the head). Queues yield items in a first-in, first-out or FIFO fashion. Items are removed in the same order in which they were added.
	队列添加元素是添加在队尾上，删除元素是从队列的头部开始。队列的特点先进先出。元素的添加和删除有相同的顺序。

```
Queue<String> q = new LinkedList<String>();
```

```
        q.add("A"); q.add("B"); q.add("C");
        while (q.size() > 0) { System.out.print(q.remove() + " "); } // Prints
A B C
```

3）	Priority Queues（优先队列）
	A priority queue collects elements, each of which has a priority. A typical example of a priority queue is a collection of work requests, some of which may be more urgent than others. Unlike a regular queue, the priority queue does not maintain a first-in, first-out discipline. Instead, elements are retrieved according to their priority. In other words, new items can be inserted in any order. But whenever an item is removed, it is the item with the most urgent priority.
	优先队列里面的元素有优先级。一个典型的例子是，在很多工作请求中有些是十分紧急的，而其他的并不紧急。跟普通队列不同的是优先队列并不遵守先进先出的约定。在优先队列中，元素的出队根据元素的优先级而定。换句话说，新的元素可以添加在任何位置。但是当出队时，优先级最高的先出队。
	It is customary to give low values to urgent priorities, with priority 1 denoting the most urgent priority. Thus, each removal operation extracts the minimum element from the queue.
	习惯上是给最高优先级一个比较小的数字。用数字 1 代表最高的优先级。因此，优先级数字最小的元素先出队。

```
        PriorityQueue<Integer> q =new PriorityQueue<Integer>();
        q.add(3); q.add(1); q.add(2);
        int first = q.remove();//first is set to 1
        int second = q.remove();//second to 2.
        int next = q.peek();//Gets the smallest value in the priority queue without
removing it.
```

7. Choosing a Collection（选择一个集合）

Step 1　Determine how you access the values;
Step 2　Determine the element types or key/value types;
Step 3　Determine whether element or key order matters;
Step 4　For a collection, determine which operations must be fast;
Step 5　For hash sets and maps, decide whether you need to implement the hashCode and equals methods;
Step 6　If you use a tree, decide whether to supply a comparator.

第一步：确定你要存储的数据；
第二步：确定这个元素类型或者键/值类型；
第三步：确定元素或者键的顺序是否重要；
第四步：对于一个集合而言，确定哪一个操作是首选；
第五步：对于 hash 集合和 map 集合，确定你是否需要实现 hashCode 和 equals 方法；
第六步：如果使用树，确定是否提供一个 comparator。

9.2 Example（例题）

<table>
<tr><td rowspan="2">1.</td><td>Consider how you write arithmetic expressions, such as (3+4)*5. The parentheses are needed so that 3 and 4 are added before multiplying the result by 5.However, you can eliminate the parentheses if you write the operators after the numbers, like this: 3 4 + 5*. To evaluate this expression, apply + to 3 and 4, yielding 7, and then simplify 7 5* to 35. We put it on a stack.</td></tr>
<tr><td>考虑如何求一个算术表达式，例如（3+4）*5。这个括号的意思是 3 加 4 的操作是在乘以 5 之前进行的。然而，如果把操作符写在数字后面可以省略括号，就像：3 4+5*。提供此表达式，用+连接 3 和 4，得到 7，同样的 7 5*得到 35。把它放入一个栈内。</td></tr>
</table>

```
import java.util.Scanner;
import java.util.Stack;
/**
   This calculator uses the reverse Polish notation.
*/
public class Calculator
{
   public static void main(String[] args)
   {
      Scanner in = new Scanner(System.in);
      Stack<Integer> results = new Stack<Integer>();
      System.out.println("Enter one number or operator per line, Q to quit. ");
      boolean done = false;
      while (!done)
      {
         String input = in.nextLine();
   // If the command is an operator, pop the arguments and push the result
         if (input.equals("+"))
         {
            results.push(results.pop() + results.pop());
         }
         else if (input.equals("-"))
         {
            Integer arg2 = results.pop();
            results.push(results.pop() - arg2);
         }
         else if (input.equals("*") || input.equals("x"))
         {
            results.push(results.pop() * results.pop());
         }
         else if (input.equals("/"))
         {
```

```
              Integer arg2 = results.pop();
              results.push(results.pop() / arg2);
            }
            else if (input.equals("Q") || input.equals("q"))
            {
              done = true;
            }
            else
            {
              // Not an operator--push the input value
              results.push(Integer.parseInt(input));
            }
            System.out.println(results);
          }
        }
      }
    Running for like this:
```

应用栈的结果如图 9-8 所示。

```
Enter one number or operator per line, Q to quit.
3
[3]
4
[3, 4]
+
[7]
5
[7, 5]
 *
[35]
```

Fig9-8.　Using Stack

2.	The following program shows a practical application of sets. It reads in all words from a dictionary file that contains correctly spelled words and places them in a set. It then reads all words from a document—here, the book Alice in Wonderland—into a second set. Finally, it prints all words from that set that are not in the dictionary set. These are the potential misspellings. (As you can see from the output, we used an American dictionary, and words with British spelling, such as clamour, are flagged as potential errors.) 接下来的例子演示的是一个 set 集合的使用。它从一个字典文件里读取了所有的单词，其中包含了正确的拼写并把它们放入一个 set 集合。接着把另一个文档里的单词（比如，爱丽丝历险记）放在另外一个集合里。输出第二个集合中第一个集合不包含的单词。可能会出现拼写不准的单词（我们用美国的字典，然后使用英式拼写，例如，clamour，就会被标为拼写错误）。

```
import java.util.HashSet;
import java.util.Scanner;
import java.util.Set;
import java.io.File;
```

```
import java.io.FileNotFoundException;
/**
  This program checks which words in a file are not present in a dictionary.
*/
public class SpellCheck
{
  public static void main(String[] args)
    throws FileNotFoundException
  {
    // Read the dictionary and the document
    Set<String> dictionaryWords = readWords("words");
    Set<String> documentWords = readWords("alice30.txt");
    // Print all words that are in the document but not the dictionary
    for (String word : documentWords)
    {
      if (!dictionaryWords.contains(word))
      {
        System.out.println(word);
      }
    }
  }
  /**
    Reads all words from a file.
    @param filename the name of the file
    @return a set with all lowercased words in the file. Here, a
    word is a sequence of upper- and lowercase letters.
  */
  public static Set<String> readWords(String filename)
    throws FileNotFoundException
  {
    Set<String> words = new HashSet<String>();
    Scanner in = new Scanner(new File(filename));
    // Use any characters other than a-z or A-Z as delimiters
    in.useDelimiter("[^a-zA-Z]+");
    while (in.hasNext())
    {
      words.add(in.next().toLowerCase());
    }
    return words;
  }
}
```

9.3 Experimental contents（实验内容）

1.	Build a class called LinkedListRunner with a main method that instantiates a LinkedList<String>. Add the following strings to the linked list:

新建一个名字叫 LinkedListRunner 的类，其中包含一个主函数一个 LinkedList<String>实例，在链接表中加入如下的字符串：

```
aaa
bbb
ccc
ddd
eee
fff
ggg
hhh
iii
```

Build a ListIterator<String> and use it to walk sequentially through the linked list using hasNext and next, printing each string that is encountered. When you have printed all the strings in the list, use the hasPrevious and previous methods to walk backwards through the list. Along the way, examine each string and remove all the strings that begin with a vowel. When you arrive at the beginning of the list, use hasNext and next to go forward again, printing out each string that remains in the linked list.

建立一个 ListIterator<String>，用 hashNext 和 next 遍历并输出链表中的每一个元素，然后使用 hasPrevious 和 previous 方法反向输出链表，删除以元音开头的字符串，然后用 hashNext 和 next 再遍历一次并输出链表中剩下的字符串。

2. Write a program that stores the following information in a HashMap:

新建一个程序，存储如下的数据到 hashMap 中：

```
Sue is friends with Bob, Jose, Alex, and Cathy
Cathy is friends with Bob and Alex
Bob is friends with Alex, Jose, and Jerry
```

After storing the information, prompt the user to enter a name. If the name that is entered is Sue, Cathy, or Bob, print out the name and the list of friends. Otherwise print a message indicating that the name is not in the HashMap.

存储完数据之后，允许用户输入一个名字，如果这个名字在这个 HashMap 中，则输出这个名字以及名字对应的朋友，如果不存在就输出名字不在 HashMap 中。

9.4 Experimental steps（实验步骤）

1.

```
import java.util.LinkedList;
import java.util.ListIterator;
public class LinkedListRunner  {
   public static void main(String[] args) {
       // TODO Auto-generated method stub
```

```
        LinkedList<String> linkedList=new LinkedList<String>();
        linkedList.add("aaa");
        linkedList.add("bbb");
        linkedList.add("ccc");
        linkedList.add("ddd");
        linkedList.add("eee");
        linkedList.add("fff");
        linkedList.add("ggg");
        linkedList.add("hhh");
        linkedList.add("iii");
        ListIterator<String> iterator=linkedList.listIterator();
        while(iterator.hasNext())
        {
            System.out.println(iterator.next());
        }
        while(iterator.hasPrevious())
        {
            char c=iterator.previous().charAt(0);
            if(c=='a'||c=='e'||c=='i'||c=='o'||c=='u')
            {
                iterator.remove();
            }
        }
        System.out.println("**********");
        while(iterator.hasNext())
        {
            System.out.println(iterator.next());
        }
    }
}
```

2.

```
import java.util.HashMap;
import java.util.Scanner;
public class UseHashMap {
    public static void main(String[] args) {
        // TODO Auto-generated method stub
        HashMap<String,String> hashMap=new HashMap<String,String>();
        hashMap.put("Sue", "Bob, Jose, Alex, and Cathy");
        hashMap.put("Cathy", "Bob and Alex");
        hashMap.put("Bob", "Alex, Jose, and Jerry");
        Scanner in=new Scanner(System.in);
        String str=in.next();
        in.close();
        if(hashMap.keySet().contains(str))
        {
            System.out.println(str+" is friends with "+hashMap.get(str));
        }
```

```
            else
            {
                System.out.println("There is no this key!");
            }
        }
    }
```

9.5　Experimental result（实验结果）

实验结果如图 9-9 和图 9-11 所示。

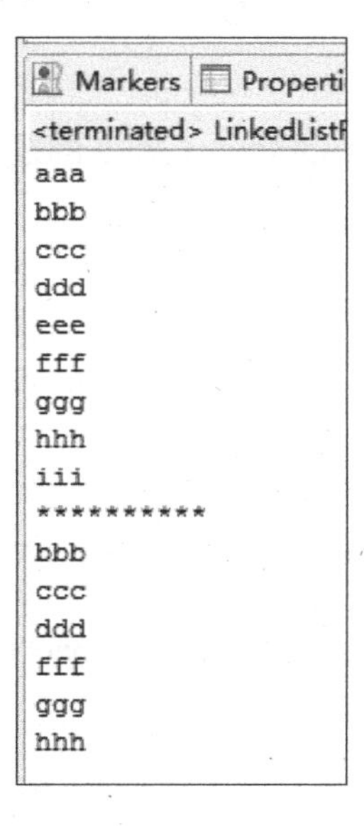

Fig9-9.　the Result of Using LinkedList

Markers | Properties | Servers | Data Source Explore

<terminated> UseHashMap [Java Application] C:\Program Files\J

Sue

Sue is friends with Bob, Jose, Alex, and Cathy

Fig9-10.　the Result 1 of Using HashMap

<terminated> UseHashMap [Java Application] C:\

arbot

There is no this key!

Fig9-11.　the Result 2 of Using HashMap

第 10 章　Streams and Binary Input/Output（流与二进制的输入/输出）

10.1　Key points of this chapter（本章要点）

1. Java Classes for Input and Output（Java 的输入和输出类）

Java 的输入/输出类如图 10-1 所示。

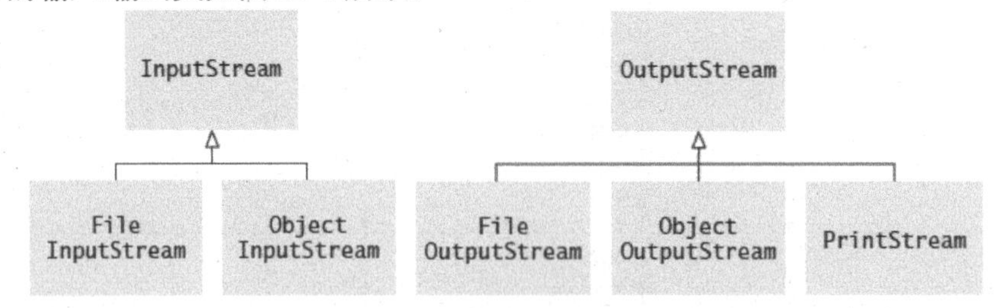

Fig10-1.　Java Classes for Input and Output

Java 的读取和写入类如图 10-2 所示。

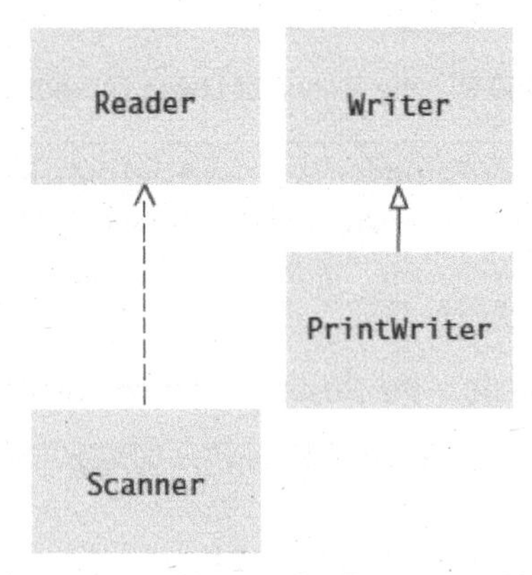

Fig10-2.　Java Classes for Reader and Writer

2. Text Data（文本数据）

Reader and Writer and their subclasses were designed to process text input and output.
Reader 和 Writer 类以及它们的子类被用来处理文本的输入/输出。

```
Scanner in = new Scanner(input, "UTF-8");
   // Input can be a File or InputStream
```

```
PrintWriter out = new PrintWriter(output, "UTF-8");
  // Output can be a File or OutputStream
```

3. Binary Input and Output（二进制的输入和输出）

Use InputStream and OutputStream and their subclasses to process binary input and output.
使用 InputStream 和 OutputStream 及它们的子类用来处理二进制的输入和输出。

```
FileInputStream inputStream =new FileInputStream("input.bin");
FileOutputStream outputStream =new FileOutputStream("output.bin");
```

1）	Binary Input（二进制输入）
	Use read method of InputStream class to read a single byte.
	使用 InputStream 类的 read 方法来读取一个字节。

```
InputStream in = . . .;
int next = in.read();
if (next != -1)
{
  Process next // a value between 0 and 255
}
```

2）	Binary Output（二进制输出）
	Use write method of OutputStream class to write a single byte.
	使用 OutputStream 类的 write 方法来写入一个字节。

```
OutputStream out = . . .;
int value= . . .; // should be between 0 and 255
out.write(value);
```

3. Random Access（随机存取）

RandomAccessFile Class（RandomAccessFile 类）	
	Open a file with open mode:Reading only ("r"),Reading and writing ("rw").
	以某种“打开方式”打开一个文件：只读方式为（“r”），读写方式(“rw”)。
	RandomAccessFile f =new RandomAcessFile("bank.dat","rw"); To move the file pointer to a specific byte: f.seek(position);
	移动文件的指针到指定的字节；
	To get the current position of the file pointer:
	得到当前文件指针的位置：

```
long position = f.getFilePointer();// of type "long" because files can
be very large //使用 long 类型是因为文件长度可能很大
```

	To find the number of bytes in a file:
	得到文件字节的数量。

```
long fileLength = f.length();
```

4. Object Streams（对象流）

ObjectOutputStream class can save entire objects to disk.

ObjectOutputStream 类可以存储所有的对象到磁盘中。

ObjectInputStream class can read them back in.

ObjectInputStream 类可以把对象从磁盘读取出来。

1） Writing an Object to File（写一个对象到文件中）

```
BankAccount b = ...;
ObjectOutputStream out = new ObjectOutputStream(
  new FileOutputStream("bank.dat"));
out.writeObject(b);
2)Reading an Object From File（从文件中读取一个对象）
ObjectInputStream in = new ObjectInputStream(
  new FileInputStream("bank.dat"));
BankAccount b =(BankAccount) in.readObject();
```

2） Serializable Interface（序列化接口）

Objects that are written to an object stream must belong to a class that implements the Serializable interface:

被写入对象流的对象必须属于一个已经实现 Serializable 接口的类。

```
class BankAccount implements Serializable
{
  …
}
```

10.2 Example（例题）

Following is a sample program that puts serialization to work. The Bank class manages a collection of bank accounts. Both the Bank and BankAccount classes implement the Serializable interface. Run the program several times. Whenever the program exits, it saves the Bank object (and all bank account objects that the bank contains) into a file bank.dat. When the program starts again, the file is loaded, and the changes from the preceding program run are automatically reflected. However, if the file is missing (either because the program is running for the first time, or because the file was erased), then the program starts with a new bank.

下面是一个把序列化应用到工作中的示范程序。Bank 类管理 BankAccount 的集合。Bank 和 BankAccount 类都实现了 Serializable 接口。运行程序几次。无论程序在何时退出，它都会把 Bank 对象（以及所有 Bank 包含的 BankAccount 对象）存入 bank.dat 文件。当程序再次开始执行，文件就会被载入，先前程序运行的变化会自动反映出来。然而，如果文件不存在（可能因为程序是第一次运行，或者文件被清除了），程序就会新建一个 bank。

```
//Bank.java
import java.io.Serializable;
import java.util.ArrayList;
/**
   This bank contains a collection of bank accounts.
*/
public class Bank implements Serializable
{
   private ArrayList<BankAccount> accounts;
   /**
      Constructs a bank with no bank accounts.
         */
   public Bank()
   {
      accounts = new ArrayList<BankAccount>();
   }
   /**
      Adds an account to this bank.
      @param a the account to add

   */
   public void addAccount(BankAccount a)
   {
      accounts.add(a);
   }
   /**
      Finds a bank account with a given number.
      @param accountNumber the number to find
      @return the account with the given number, or null if there
      is no such account
   */
   public BankAccount find(int accountNumber)
   {
      for (BankAccount a : accounts)
      {
         if (a.getAccountNumber() == accountNumber) // Found a match
         {
            return a;
         }
      }
      return null; // No match in the entire array list
   }
}
```

```
//BankAccount.java
import java.io.Serializable;
/**
   A bank account has a balance that can be changed by
   deposits and withdrawals.
*/
public class BankAccount implements Serializable
{
   private int accountNumber;
   private double balance;
   /**
      Constructs a bank account with a zero balance.
      @param anAccountNumber the account number for this account
   */
   public BankAccount(int anAccountNumber)
   {
      accountNumber = anAccountNumber;
      balance = 0;
   }
   /**
      Constructs a bank account with a given balance.
      @param anAccountNumber the account number for this account
      @param initialBalance the initial balance
   */
   public BankAccount(int anAccountNumber, double initialBalance)
   {
      accountNumber = anAccountNumber;
      balance = initialBalance;
   }
   /**
      Gets the account number of this bank account.
      @return the account number
   */
   public int getAccountNumber()
   {
      return accountNumber;
   }
   /**
      Deposits money into the bank account.
      @param amount the amount to deposit
   */
   public void deposit(double amount)
   {
      double newBalance = balance + amount;
      balance = newBalance;
   }
   /**
```

```
      Withdraws money from the bank account.
      @param amount the amount to withdraw
    */
    public void withdraw(double amount)
    {
      double newBalance = balance - amount;
      balance = newBalance;
    }
    /**
      Gets the current balance of the bank account.
      @return the current balance
    */
    public double getBalance()
    {
      return balance;
    }
  }
  //SerialDemo.java
  import java.io.File;
  import java.io.IOException;
  import java.io.FileInputStream;
  import java.io.FileOutputStream;
  import java.io.ObjectInputStream;
  import java.io.ObjectOutputStream;
  /**
    This program demonstrates serialization of a Bank object.
    If a file with serialized data exists, then it is loaded.
    Otherwise the program starts with a new bank.
    Bank accounts are added to the bank. Then the bank
    object is saved.
  */
  public class SerialDemo
  {
    public static void main(String[] args)
        throws IOException, ClassNotFoundException
    {
      Bank firstBankOfJava;
      File f = new File("bank.dat");
      if (f.exists())
      {
        ObjectInputStream in = new ObjectInputStream(
          new FileInputStream(f));
        firstBankOfJava = (Bank) in.readObject();
        in.close();
      }
      else
      {
```

```
            firstBankOfJava = new Bank();
            firstBankOfJava.addAccount(new BankAccount(1001, 20000));
            firstBankOfJava.addAccount(new BankAccount(1015, 10000));
        }
        // Deposit some money
        BankAccount a = firstBankOfJava.find(1001);
        a.deposit(100);
        System.out.println(a.getAccountNumber() + ": " + a.getBalance());
        a = firstBankOfJava.find(1015);
        System.out.println(a.getAccountNumber() + ": " + a.getBalance());
        ObjectOutputStream out = new ObjectOutputStream(
            new FileOutputStream(f));
        out.writeObject(firstBankOfJava);
        out.close();
    }
}
```

10.3 Experimental contents（实验内容）

1.	You will write a program that writes memos to a file as object streams. Your program will allow a user to enter multiple memos, supplying the topic and message. First, create a Memo object. The Memo object should have instance variables to store a topic, a date stamp, and the memo message. Remember to implement the Serializable interface! 编写一个程序把备忘录写入文件应用对象流。程序必须允许使用者输入多行备忘信息，包括主题和内容。首先，新建一个 Memo 对象，Memo 对象中要有存储主题、日期、备忘内容的实例化变量，并且还要实现 Serializable 接口！

2.	Now provide a memo manager class that writes Memo objects to a file as an object stream. Allow a user to enter multiple memos, supplying the topic and message. Your program should provide the date stamp using the java.util.Date object (creating a Date object with no arguments initializes the object to the current time and date). Let the user choose the file name to which the memos are saved. After saving the file, your memo manager program should open the file and print out all memos. 编写一个备忘录管理类，用来把备忘录作为一个对象流写入文件中。允许使用者可以输入多行备忘信息，包括主题和内容，程序应该提供日期标志，可以使用 java.util.Date 对象（创建一个不带参数的 Date 对象，表示当前的时间和日期）。让使用者自己选择备忘录的文件名。存储该文件后，备忘录管理程序打开该文件并输出所有备忘录。

10.4 Experimental steps（实验步骤）

1.

```
import java.io.Serializable;
public class Memo implements Serializable
{
  private String topic;
  private String dateStamp;
  private String message;
  public Memo(String aTopic, String aDateStamp, String aMessage)
  {
    topic = aTopic;
    dateStamp = aDateStamp;
    message = aMessage;
  }

  public String getTopic()
  {
    return topic;
  }
  public String getDateStamp()
  {
    return dateStamp;
  }
  public String getMessage()
  {
    return message;
  }
}
```

2.

```
import java.util.Date;
import java.io.IOException;
import java.io.ObjectOutputStream;
import java.io.FileOutputStream;
import java.io.ObjectInputStream;
import java.io.FileInputStream;
import java.util.Scanner;
import java.util.ArrayList;
public class MemoManager
{
  public static void main(String[] args)
  {
    Scanner console = new Scanner(System.in);
    System.out.println("Output file:");
    String filename = console.nextLine();
    ArrayList<Memo> memos = new ArrayList<Memo>();
```

```
        try
        {
          ObjectOutputStream out = new ObjectOutputStream(
            new FileOutputStream(filename));
          boolean done = false;
          while (!done)
          {
            System.out.println("Memo topic (enter -1 to end):");
            String topic = console.nextLine();
            if (topic.equals("-1"))
            {
              done = true;
            }
            else
            {
              System.out.println("Memo text:");
              String message = console.nextLine();
              String dateStamp = (new Date()).toString();
              Memo memo = new Memo(topic, dateStamp, message);
              memos.add(memo);
            }
          }
          out.writeObject(memos);
          out.close();
          // Open the file and print all the memos
          ObjectInputStream in = new ObjectInputStream(new
            FileInputStream(filename));
          ArrayList<Memo> savedMemos = (ArrayList<Memo>) in.readObject();
          for (Memo memo : savedMemos)
          {
            System.out.println(memo.getTopic() + "\n" +
              memo.getDateStamp() + "\n" + memo.getMessage());
          }
        }
        catch (IOException exception)
        {
          System.out.println("Error processing file: " + exception);
        }
        catch (ClassNotFoundException exception)
        {
          System.out.println("Error casting class: " + exception);
        }
    }
}
```

10.5 Experimental result（实验结果）

实验结果如图 10-3 所示。

```
Output file:
d:\\test1.txt
Memo topic (enter -1 to end):
hello
Memo text:
hello
Memo topic (enter -1 to end):
world
Memo text:
world
Memo topic (enter -1 to end):
-1
hello
Sun Nov 13 21:44:18 CST 2016
hello
world
Sun Nov 13 21:44:28 CST 2016
world
```

Fig10-3. Result of Experiment

第 11 章　Multithreading（多线程）

11.1　Key points of this chapter（本章要点）

1．Running Threads（运行线程）

1）	Implement a class that implements the Runnable interface（实现一个 Runnable 接口的类）

```
public interface Runnable
{
   void run();
}
```

2）	Place the code for your task into the run method of your class（将此任务代码放入到你的类的 run()方法中）

```
public class MyRunnable implements Runnable
{
   public void run()
   {
      Task statements
      . . .
   }
}
```

3）	Create an object of your subclass（创建一个子类对象）

```
Runnable r = new MyRunnable();
```

4）	Construct a Thread object from the Runnable object（对 Runnable 对象构建一个线程对象）

```
Thread t = new Thread(r);
```

5）	Call the start method to start the thread（调用 start()方法启动线程）

```
t.start();
```

2．Terminating Threads(终止线程)

notify a thread that it should terminate.

通知一个线程终止。

```
t.interrupt();
```

3. Synchronizing Object Access（同步对象的访问）

To solve problems such as Object access conflict, use a lock object. Lock object is used to control threads that manipulate shared resources. ReentrantLock is most commonly used lock class.

为了解决对象访问冲突的问题，使用一个 Lock 对象。Lock 对象用来控制处理共享资源的线程。ReentrantLock 是最常用的 lock 类。

Typically, a Lock object is added to a class whose methods access shared resources, like this:

通常情况下，一个 Lock 对象会被添加到一个其方法访问了共享资源的类中，如下：

```
public class BankAccount
{
   private Lock balanceChangeLock;
   . . .
   public BankAccount()
   {
      balanceChangeLock = new ReentrantLock();
      . . .
   }
   . . .
}
```

Code that manipulates shared resource is surrounded by calls to lock and unlock, like this:

为了实现对共享资源的使用，通过调用 lock()和 unlock()方法来实现，如下所示：

```
balanceChangeLock.lock();
try
{
   Manipulate the shared resource.
}
finally
{
   balanceChangeLock.unlock();
}
```

4. Avoiding Deadlocks（避免死锁）

A deadlock occurs if no thread can proceed because each thread is waiting for another to do some work first. To overcome this problem, use a condition object, Condition objects allow a thread to temporarily release a lock, and to regain the lock at a later time. Each condition object belongs to a specific lock object. You can obtain a condition object with newCondition method of Lock interface，like this:

如果没有线程可以继续，死锁就会发生，因为每个线程都在等待另一个线程来做一些工作。要克服这一问题，可以使用 condition 对象，condition 对象允许线程暂时释放 lock，并在稍后重新获取它。每个 condition 对象属于一个特定的 lock 对象。你可以用 newCondition 方法的 lock 接口获得 condition 对象，如下所示：

```
public class BankAccount
{
  private Lock balanceChangeLock;
  private Condition sufficientFundsCondition;
  . . .
  public BankAccount()
  {
    balanceChangeLock = new ReentrantLock();
    sufficientFundsCondition =balanceChangeLock.newCondition();
    . . .
  }
}
```

It is customary to give the condition object a name that describes the condition that you want to test (such as “sufficient funds”). You need to implement an appropriate test. For as long as the test is not fulfilled, call the await method on the condition object, like this:

习惯上给 condition 对象一个描述你想要测试的条件的名称（如“足够的资金”）。你需要实施一个适当的测试。只要测试没有完成，就调用 condition 对象的 await 方法，如下所示：

```
public void withdraw(double amount)
{
   balanceChangeLock.lock();
   try
   {
   while (balance < amount)
   {
   sufficientFundsCondition.await();
   }
   . . .
   }
   finally
   {
   balanceChangeLock.unlock();
   }
}
```

When a thread calls await, it is not simply deactivated in the same way as a thread that reaches the end of its time slice. Instead, it is in a blocked state, and it will not be activated by the thread scheduler until it is unblocked. To unblock, another thread must execute the signalAll method on the same condition object. The signalAll method unblocks all threads waiting on the condition. They can then compete with all other threads that are waiting for the lock object. Eventually, one of them will gain access to the lock, and it will exit from the await method，like this:

当一个线程调用 await 时，它并不是简单地以像到达时间片结束的线程那样去激活。相反，它处于阻塞状态，直到其畅通它才会被线程程序激活。为了解除阻塞，另一个线程必须对相同的 condition 对象执行 signalAll 方法。signalAll 方法将解锁所有等待的线程。然后，它们可以与其他等待锁定对象的线程竞争。最终，它们中的一个将获得 lock，并且它将退出 await 方法。如下所示：

```
public void deposit(double amount)
{
   balanceChangeLock.lock();
   try
   {
   . . .
   sufficientFundsCondition.signalAll();
   }
   finally
   {
   balanceChangeLock.unlock();
   }
}
```

11.2 Example（例题）

We construct a bank account that starts out with a zero balance. We create two sets of threads: Each thread in the first set repeatedly deposits $100.Each thread in the second set repeatedly withdraws $100. With the calls to await and signalAll in the withdraw and deposit methods, we can launch any number of withdrawal and deposit threads without a deadlock. If you run the sample program, you will note that all transactions are carried out without ever reaching a negative balance.

我们设立一个余额为零的银行账户。创造了两个线程：第一个线程实现为银行账户中反复存入 $100。第二个线程实现为银行账户中反复提取 $100。提款和存款时调用 await 和 signalAll 方法，我们可以不出现死锁地启动提款和存款线程。如果运行示例程序，你会注意到，所有的交易都是在未出现负平衡的时候进行的。

```
// BankAccount.java
import java.util.concurrent.locks.Condition;
import java.util.concurrent.locks.Lock;
import java.util.concurrent.locks.ReentrantLock;
/**
```

```
   A bank account has a balance that can be changed by
   deposits and withdrawals.
*/
public class BankAccount
{
   private double balance;
   private Lock balanceChangeLock;
   private Condition sufficientFundsCondition;
   /**
      Constructs a bank account with a zero balance.
   */
   public BankAccount()
   {
      balance = 0;
      balanceChangeLock = new ReentrantLock();
      sufficientFundsCondition = balanceChangeLock.newCondition();
   }
   /**
      Deposits money into the bank account.
      @param amount the amount to deposit
   */
   public void deposit(double amount)
   {
      balanceChangeLock.lock();
      try
      {
         System.out.print("Depositing " + amount);
         double newBalance = balance + amount;
         System.out.println(", new balance is " + newBalance);
         balance = newBalance;
         sufficientFundsCondition.signalAll();
      }
      finally
      {
         balanceChangeLock.unlock();
      }
   }
   /**
      Withdraws money from the bank account.
      @param amount the amount to withdraw
   */
   public void withdraw(double amount) throws InterruptedException
   {
      balanceChangeLock.lock();
      try
      {
         while (balance < amount)
```

```
         {
            sufficientFundsCondition.await();
         }
         System.out.print("Withdrawing " + amount);
         double newBalance = balance - amount;
         System.out.println(", new balance is " + newBalance);
         balance = newBalance;
      }
      finally
      {
         balanceChangeLock.unlock();
      }
   }
   /**
      Gets the current balance of the bank account.
      @return the current balance
   */
   public double getBalance()
   {
      return balance;
   }
}
// DepositRunnable.java
/**
   A deposit runnable makes periodic deposits to a bank account.
*/
public class DepositRunnable implements Runnable
{
   private static final int DELAY = 1;
   private BankAccount account;
   private double amount;
   private int count;
   /**
      Constructs a deposit runnable.
      @param anAccount the account into which to deposit money
      @param anAmount the amount to deposit in each repetition
      @param aCount the number of repetitions
   */
   public DepositRunnable(BankAccount anAccount, double anAmount,
         int aCount)
   {
      account = anAccount;
      amount = anAmount;
      count = aCount;
   }
   public void run()
   {
```

```
      try
      {
         for (int i = 1; i <= count; i++)
         {
            account.deposit(amount);
            Thread.sleep(DELAY);
         }
      }
      catch (InterruptedException exception) {}
   }
}
// WithdrawRunnable.java
/**
   A withdraw runnable makes periodic withdrawals from a bank account.
*/
public class WithdrawRunnable implements Runnable
{
   private static final int DELAY = 1;
   private BankAccount account;
   private double amount;
   private int count;
   /**
      Constructs a withdraw runnable.
      @param anAccount the account from which to withdraw money
      @param anAmount the amount to withdraw in each repetition
      @param aCount the number of repetitions
   */
   public WithdrawRunnable(BankAccount anAccount, double anAmount,
         int aCount)
   {
      account = anAccount;
      amount = anAmount;
      count = aCount;
   }
   public void run()
   {
      try
      {
         for (int i = 1; i <= count; i++)
         {
            account.withdraw(amount);
            Thread.sleep(DELAY);
         }
      }
      catch (InterruptedException exception) {}
   }
}
```

```
// BankAccountThreadRunner.java
/**
   This program runs threads that deposit and withdraw
   money from the same bank account.
*/
public class BankAccountThreadRunner
{
   public static void main(String[] args)
   {
      BankAccount account = new BankAccount();
      final double AMOUNT = 100;
      final int REPETITIONS = 100;
      final int THREADS = 100;
      for (int i = 1; i <= THREADS; i++)
      {
         DepositRunnable d = new DepositRunnable(
            account, AMOUNT, REPETITIONS);
         WithdrawRunnable w = new WithdrawRunnable(
            account, AMOUNT, REPETITIONS);
         Thread dt = new Thread(d);
         Thread wt = new Thread(w);
         dt.start();
         wt.start();
      }
   }
}
```

11.3 Experimental contents（实验内容）

1.	Write a program WordCount that counts the words in one or more files. Start a new thread for each file. 编写一个名为 WordCount 的程序，统计一个或多个文件中的字数。为每个文件开启一个新线程。

For example, if you call java WordCount report.txt address.txt Homework.java
then the program might print:

```
address.txt: 1052
Homework.java: 445
report.txt: 2099
```

2.	Add a condition to the deposit method of the BankAccount class, restricting deposits to $100,000 (the insurance limit of the U.S. government). The method should block until sufficient money has been withdrawn by another thread. Test your program with a large number of deposit threads.

添加一条限制存款十万元（美国政府的保险限额）的条件到 BankAccount 类的 deposit 方法中。该方法将阻塞线程，直到其他线程实现了钱被充分取出。用大量的 deposit 线程来测试你的程序。

```
(BankAccount.java;WithdrawRunnable.java;BankAccountThreadRunner.java)
public class BankAccount
{
  private int accountNumber;
  private double balance;
  public BankAccount(int anAccountNumber)
  {
    accountNumber = anAccountNumber;
    balance = 0;
  }
   public BankAccount(int anAccountNumber, double initialBalance)
  {
    accountNumber = anAccountNumber;
    balance = initialBalance;
  }
  public int getAccountNumber()
  {
    return accountNumber;
  }
 public void deposit(double amount)
  {
    double newBalance = balance + amount;
    balance = newBalance;
  }
  public void withdraw(double amount)
  {
    double newBalance = balance - amount;
    balance = newBalance;
  }
 public double getBalance()
  {
    return balance;
  }
}
public class DepositRunnable implements Runnable
{
   //……
}
public class WithdrawRunnable implements Runnable
{
   //……
}
public class BankAccountThreadRunner
```

```
{
  public static void main(String[] args)
  {
//......
  }
}
```

11.4 Experimental steps（实验步骤）

1.

```
import java.io.File;
import java.io.FileNotFoundException;
import java.io.FileReader;
import java.io.IOException;
/**
 * WordCount that counts the words in one or more files
 * @author *****
 *
 */
public class WordCount  implements Runnable {
    private String filename;
    public WordCount(String filename) {
        this.filename = filename;
    }
    @Override
    public void run() {
        File file=new File("E:/",filename);
        try {
            FileReader r=new FileReader(file);
            int length=(int)file.length();
            char byte[]=new char[len];
            try {
                r.read(byte);
            } catch (IOException e) {
                // TODO Auto-generated catch block
                e.printStackTrace();
            }
            String str=String.valueOf(byte);/
            String[] str1=str.split("[.,\"\\?!:'\\s]");
            System.out.print(filename+":");
            System.out.println(str1.length);
        } catch (FileNotFoundException e) {
            // TODO Auto-generated catch block
            e.printStackTrace();
        }
    }
```

```
}
```

2.

```
import java.util.concurrent.locks.Condition;
import java.util.concurrent.locks.Lock;
import java.util.concurrent.locks.ReentrantLock;
/**
  A bank account has a balance that can be changed by
  deposits and withdrawals.
*/
public class BankAccount
{
  public static final double MAX_BALANCE = 100000;
  private double balance;
  private Lock balanceChangeLock;
  private Condition sufficientFundsCondition;
  private Condition lessThanMaxBalanceCondition;
  /**
    Constructs a bank account with a zero balance.
  */
  public BankAccount()
  {
    balance = 0;
    balanceChangeLock = new ReentrantLock();
    sufficientFundsCondition = balanceChangeLock.newCondition();
    lessThanMaxBalanceCondition = balanceChangeLock.newCondition();
  }
  /**
    Deposits money into the bank account.
    @param amount the amount to deposit
  */
  public void deposit(double amount)
      throws InterruptedException
  {
    balanceChangeLock.lock();
    try
    {
      while (balance + amount > MAX_BALANCE)
        lessThanMaxBalanceCondition.await();
      System.out.print("Depositing " + amount);
      double newBalance = balance + amount;
      System.out.println(", new balance is " + newBalance);
      balance = newBalance;
      sufficientFundsCondition.signalAll();
    }
    finally
    {
```

```
            balanceChangeLock.unlock();
         }
      }
      /**
         Withdraws money from the bank account.
         @param amount the amount to withdraw
      */
      public void withdraw(double amount)
            throws InterruptedException
      {
         balanceChangeLock.lock();
         try
         {
            while (balance < amount)
               sufficientFundsCondition.await();
            System.out.print("Withdrawing " + amount);
            double newBalance = balance - amount;
            System.out.println(", new balance is " + newBalance);
            balance = newBalance;
            lessThanMaxBalanceCondition.signalAll();
         }
         finally
         {
            balanceChangeLock.unlock();
         }
      }
      /**
         Gets the current balance of the bank account.
         @return the current balance
      */
      public double getBalance()
      {
         return balance;
      }
   }
   /**
      A deposit runnable makes periodic deposits to a bank account.
   */
   public class DepositRunnable implements Runnable
   {
      private static final int DELAY = 1;
      private BankAccount account;
      private double amount;
      private int count;
      /**
         Constructs a deposit runnable.
```

```
      @param anAccount the account into which to deposit money
      @param anAmount the amount to deposit in each repetition
      @param aCount the number of repetitions
   */
   public DepositRunnable(BankAccount anAccount, double anAmount,
         int aCount)
   {
      account = anAccount;
      amount = anAmount;
      count = aCount;
   }
   public void run()
   {
      try
      {
         for (int i = 1; i <= count; i++)
         {
            account.deposit(amount);
            Thread.sleep(DELAY);
         }
      }
      catch (InterruptedException exception) {}
   }
}
/**
   A deposit runnable makes periodic deposits to a bank account.
*/
public class DepositRunnable implements Runnable
{
   private static final int DELAY = 1;
   private BankAccount account;
   private double amount;
   private int count;
   /**
      Constructs a deposit runnable.
      @param anAccount the account into which to deposit money
      @param anAmount the amount to deposit in each repetition
      @param aCount the number of repetitions
   */
   public DepositRunnable(BankAccount anAccount, double anAmount,
         int aCount)
   {
      account = anAccount;
      amount = anAmount;
      count = aCount;
   }
```

```
   public void run()
   {
      try
      {
         for (int i = 1; i <= count; i++)
         {
            account.deposit(amount);
            Thread.sleep(DELAY);
         }
      }
      catch (InterruptedException exception) {}
   }
}
/**
   This program runs threads that deposit and withdraw money from the same
     bank account.
*/
public class BankAccountThreadRunner
{
   public static void main(String[] args)
   {
      BankAccount account = new BankAccount();
      final double AMOUNT = 10000;
      final int REPETITIONS = 10;
      final int DEPOSIT_THREADS = 10;
      final int WITHDRAW_THREADS = 2;
      for (int i = 0; i < DEPOSIT_THREADS; i++)
      {
         DepositRunnable d = new DepositRunnable(account, AMOUNT,
           REPETITIONS);
         Thread t = new Thread(d);
         t.start();
      }
      for (int i = 0; i < WITHDRAW_THREADS; i++)
      {
         WithdrawRunnable d = new WithdrawRunnable(account, AMOUNT,
           REPETITIONS * DEPOSIT_THREADS / WITHDRAW_THREADS);
         Thread t = new Thread(d);
         t.start();
      }
   }
}
```

11.5 Experimental result（实验结果）

实验结果如图 11-1 和图 11-2 所示。

```
home.txt:8
address.txt:5
report.txt:4
```

Fig11-1. Count the words in one or more files

```
Withdrawing 10000.0, new balance is 60000.0
Withdrawing 10000.0, new balance is 50000.0
Withdrawing 10000.0, new balance is 40000.0
Withdrawing 10000.0, new balance is 30000.0
Withdrawing 10000.0, new balance is 20000.0
Withdrawing 10000.0, new balance is 10000.0
Withdrawing 10000.0, new balance is 0.0
```

Fig11-2. BankAccount

第 12 章　Programming with JDBC（JDBC 编程）

12.1　Key points of this chapter（本章要点）

1. JDBC: Java Database Connectivity（Java 数据库连接）

You need a JDBC driver to access a database from a Java program, Different databases require different drivers. Drivers may be supplied by the database manufacturer or a third party. When your Java program issues SQL commands, the driver forwards them to the database and lets your program analyze the results.
你需要一个 JDBC 驱动程序来访问一个 java 程序数据库，不同的数据库需要不同的驱动程序。驱动程序是由数据库制造商或第三方提供的。当 Java 程序发出 SQL 命令，驱动程序把它们转发给数据库并且使程序分析结果。

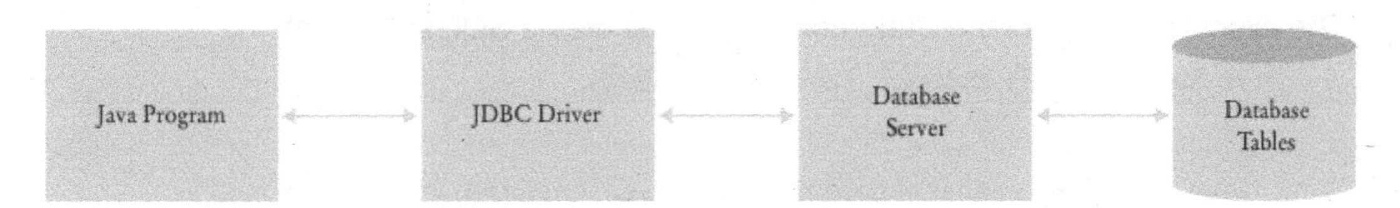

Fig12-1. JDBC Architecture（体系）

2. Database Programming in Java（用 Java 进行数据库编程）

1）	Connecting to the Database（连接数据库）
	With older versions of the JDBC standard, you first need to manually load the database driver class. Starting with JDBC4 (which is a part of Java 6), the driver is loaded automatically. If you use Java 6 or later and a fully JDBC4 compatible driver, you can skip the loading step. Otherwise, use the following code:
	对于旧版本的 JDBC 标准，你首先需要用手动加载数据库驱动类。从 JDBC4（这是 Java 6 的一部分）开始，驱动变为自动加载。如果使用 Java 6 或更高版本或一个充分兼容 JDBC4 的驱动程序，则可跳过加载这一步。否则，使用如下代码：

```
String driver = . . .;
Class.forName(driver); // Load driver
```

To connect to a database, you need an object of the Connection class. You ask the DriverManager for a connection. You need to initialize the url, username, and password strings with the values that apply to your database:

连接数据库需要一个 Connection 类的对象来完成。若向 DriverManager 发出请求连接，则需要将应用到数据库的初始化的网址、用户名和密码字符串初始化：

```
String url = . . .;
String username = . . .;
String password = . . .;
Connection conn = DriverManager.getConnection(url, username, password);
```

When you are done issuing your database commands, close the database connection by calling the close method.

当完成数据库命令时，请通过调用 close 方法来关闭数据库连接。

```
conn.close();
```

2） Executing SQL Statements（执行 SQL 语句）

Once you have a connection, you can use it to create Statement objects. You need Statement objects to execute SQL statements.

一旦有了连接，可用它来创建 Statement 对象。需声明 Statement 对象来执行 SQL 语句。

```
Statement stat = conn.createStatement();
```

The execute method of the Statement class executes a SQL statement.

使用 Statement 类的 execute 方法执行 SQL 语句。

```
For example:
stat.execute("CREATE TABLE Test (Name CHAR(20))");
stat.execute("INSERT INTO Test VALUES ('Romeo')");
```

To issue a query, use the executeQuery method of the Statement class. The query result is returned as a ResultSet object.

使用 Statement 类的 executeQuery 方法发出一个查询。查询结果以一个 ResultSet 对象返回。

```
For example:
String query = "SELECT * FROM Test";
ResultSet result = stat.executeQuery(query);
```

For UPDATE statements, you can use the executeUpdate method. It returns the number of rows affected by the statement.

对于 UPDATE 语句，可以使用 executeUpdate 方法。它返回受该语句所影响的行数。

```
String command = "UPDATE LineItem  SET Quantity = Quantity + 1"
```

```
    + " WHERE Invoice_Number = '11731'";
    int count = stat.executeUpdate(command);
    int count = stat.executeUpdate("DELETE  FROM product WHERE number = '888'
");//delete data 删除数据
```

If your statement has variable parts, then you should use a PreparedStatement instead:

如果声明有可变化部分，那么应该使用 PreparedStatement 代替：

```
    String query = "SELECT * WHERE Account_Number = ?";
    PreparedStatement stat = conn.prepareStatement(query);
    stat.setString(1, accountNumber);
```

Finally, you can use the generic execute method to execute arbitrary SQL statements. It returns a Boolean value to indicate whether the SQL command yields a result set. If so, you can obtain it with the getResultSet method. Otherwise, you can get the update count with the getUpdateCount method.

最后，可以使用一般的执行方法来执行任意的 SQL 语句。返回一个布尔值来说明它是否将产生一个结果集的 SQL 命令。如果这样的话，可用 getResultSet 方法来获取它。否则，可用 getUpdateCount 方法获取更新计数。

```
    String command = . . .;
    boolean hasResultSet = stat.execute(command);
    if (hasResultSet)
    {
        ResultSet result = stat.getResultSet();
        . . .
    }
    else
    {
        int count = stat.getUpdateCount();
        . . .
    }
```

When you are done using a ResultSet, you should close it before issuing a new query on the same statement.

当不再使用结果集时，应该在用同一语句发出新的查询前关闭它。

```
    result.close();
```

When you are done with a Statement object, you should close it. That automatically closes the associated result set.

当不再使用 Statement 对象时，应该关闭它。那将自动关闭相关的结果集。

```
    stat.close();
```

When you close a connection, it automatically closes all statements and result sets.

当关闭一个连接时，它将自动关闭所有 Statement 和 ResultSet。

3. Analyzing Query Results（分析查询结果）

when you first get a result set from the executeQuery method, no row data are available. You need to call next to move to the first row. This appears curious, but it makes the iteration loop simple:

当第一次从 executeQuery 方法得到一个结果集，行数据是不可用的。需要调用下面的代码返回到第一行。这似乎很神奇，但它使迭代循环变得简单。

```
while (result.next())
{
Inspect column data from the current row.
}
```

For example, you can fetch the product code as：

例如，您可以获取产品代码，如下所示：

```
String productCode = result.getString(1);
or
String productCode = result.getString("Product_Code");
```

4. Result Set Metadata（结果集元数据）

When you have a result set from an unknown table, you may want to know the names of the columns. You can use the ResultSetMetaData class to find out about properties of a result set. Start by requesting the metadata object from the result set. For example：

当一个未知的表中有一个结果集时，你可能想知道列的名称。可以使用 ResultSetMetaData 类来找出有关结果集的属性。从结果集请求元数据对象开始。例如：

```
ResultSetMetaData metaData = result.getMetaData();
for (int i = 1; i <= metaData.getColumnCount(); i++)
{
    String columnName = metaData.getColumnLabel(i);
    int columnSize = metaData.getColumnDisplaySize(i);
    . . .
}
```

5. Transaction（事务）

Transaction consists of a set of SQL statements, "Transaction Processing" refers to: All SQL statements in the application to ensure to execute or none. Transaction is to ensure that the important mechanism of the database data integrity and consistency. JDBC transaction processing steps are as follows:

事务由一组 SQL 语句组成，所谓“事务处理”是指：应用程序保证事务中的 SQL 语句要么全部都执行，要么一个都不执行。事务是保证数据库中数据完整性与一致性的重要机制。JDBC 事务处理步骤如下：

（1）setAutoCommit(boolean autoCommit)方法

In order to transaction processing, must close the connection object con default Settings.

为了能进行事务处理，必须关闭连接对象 con 的默认设置。

```
con.setAutoCommit(false);
```

（2）commit()方法

Connection object con call commit () method is to make the transaction all the SQL statements to take effect

连接对象 con 调用 commit()方法就是让事务中的 SQL 语句全部生效。

```
con.commit();
```

（3）rollback()方法

Connection object con call rollback () method to cancel the transaction successful execution of SQL statements to the database data update, insert, or delete operations, withdraw the SQL data changes caused by operation, the restoration of data in the database to commit () method before the execution.

连接对象 con 调用 rollback()方法，撤消事务中成功执行过的 SQL 语句对数据库数据所做的更新、插入或删除操作，即撤消引起数据发生变化的 SQL 语句操作，将数据库中的数据恢复到 commit()方法执行之前的状态。

```
con.rollback();
```

12.2　Example（例题）

For example, We use Sql server as Database, and Suppose you have the following file:

例如，我使用 SQL Server 作为数据库，并假设有以下文件：

```
Product.sql
CREATE TABLE Product
(Product_Code CHAR(7), Description VARCHAR(40), Price DECIMAL(10, 2))
INSERT INTO Product VALUES ('116-064', 'Toaster', 24.95)
INSERT INTO Product VALUES ('257-535', 'Hair dryer', 29.95)
INSERT INTO Product VALUES ('643-119', 'Car vacuum', 19.95)

import java.sql.Connection;
import java.sql.DriverManager;
import java.sql.PreparedStatement;
import java.sql.ResultSet;
```

```
    import java.sql.SQLException;
    import java.sql.Statement;
    public class DataBase {
        public static void main(String[] args) throws ClassNotFoundException,
            SQLException {
            // TODO Auto-generated method stub
            Class.forName("com.microsoft.sqlserver.jdbc.SQLServerDriver");
            System.out.println("drive is loaded");
            Connection conn = DriverManager.getConnection("jdbc:sqlserver:
                //localhost:1433;databaseName=ProductSys","sa","sa");
            try
            {
                System.out.println("database is connected");
                Statement st=conn.createStatement();
                ResultSet rs=st.executeQuery("select * from Product");
                while(rs.next()){
                    System.out.print("Product_Code="+rs.getString(1)+"  ");
                    System.out.print("Description="+rs.getString(2)+"  ");
                    System.out.println("Price="+rs.getDouble(3));
                }
            }
            finally  {
                conn.close();
            }
        }
    }
```

从数据服务器中获取数据如图 12-2 所示。

```
drive is loaded
database is connected
Product_Code=116-064  Description=Toaster  Price=24.95
Product_Code=257-535  Description=Hair dryer  Price=29.95
Product_Code=643-119  Description=Car vacuum  Price=19.95
```

Fig12-2. Get the Data from the data Server

12.3 Experimental contents（实验内容）

1.	For example, We use Sql server as Database, and Suppose you have the following file: Product.sql (This file has been mentioned in the previous) Write a program, the user input price parameters, according to the price of the user input to find all the product price is greater than the user input price of product information.
	编写一个程序，由用户输入价格参数，根据用户输入的价格查找所有产品的价格大于用户输入价格的产品信息。

Write a program, the user input price parameters, the price of the product Toaster adjusted for the user to enter the price. 编写一个程序，由用户输入价格参数，将产品‘Toaster’的价格调整为用户输入的价格。

2.	Using the transaction, reduce the value of Price in the Product table to 'Toaster' in the Product field by 10 and add 10 to the Price attribute value of the 'Hair dryer' in the Description field. 使用事务处理，将 Product 表中 Description 字段是‘Toaster’的 Price 的值减少 10，并将减少的 10 增加到 Description 字段为‘Hair dryer’的 Price 属性值上。

12.4 Experimental steps（实验步骤）

1.

```
import java.sql.Connection;
import java.sql.DriverManager;
import java.sql.PreparedStatement;
import java.sql.ResultSet;
import java.sql.SQLException;
import java.sql.Statement;
import java.util.Scanner;
public class DataBase2 {
    public static void main(String[] args) throws ClassNotFoundException,
      SQLException {
        // TODO Auto-generated method stub
        Class.forName("com.microsoft.sqlserver.jdbc.SQLServerDriver");
        Connection conn = DriverManager.getConnection("jdbc:sqlserver://
          localhost:1433;databaseName=ProductSys","sa","sa");
        try
        {
            System.out.print("Please enter the price :");
            Scanner in = new Scanner(System.in);
            double price = in.nextDouble();
            String query = "select * from Product where Price>? ";
            PreparedStatement stat = conn.prepareStatement(query);
            stat.setDouble(1, price);
            ResultSet result = stat.executeQuery();
            while(result.next()){
                System.out.print("Product_Code="+result.getString(1)+" ");
                System.out.print("Description="+result.getString(2)+" ");
```

```
                System.out.println("Price="+result.getDouble(3));
                }
            System.out.print("Please enter the Toaster's price you want to
              change  :");
            price = in.nextDouble();
            query = "update Product  set Price=? where
               Description='Toaster' " ;
            stat = conn.prepareStatement(query);
            stat.setDouble(1, price);
            int  count = stat.executeUpdate();
        }
        catch(Exception ex){
            ex.printStackTrace();
        }
        finally   {
            conn.close();
        }
    }
}
```

2.

```
import java.sql.Connection;
import java.sql.DriverManager;
import java.sql.PreparedStatement;
import java.sql.ResultSet;
import java.sql.SQLException;
import java.sql.Statement;
import java.util.Scanner;
public class DataBase3 {
    public static void main(String[] args) throws ClassNotFoundException,
      SQLException {
        // TODO Auto-generated method stub
        Class.forName("com.microsoft.sqlserver.jdbc.SQLServerDriver");
        Connection conn=null;
        try
        {
            conn = DriverManager.getConnection("jdbc:sqlserver://
              localhost:1433;databaseName= ProductSys ","sa","sa");
            conn.setAutoCommit(false);
            String query = "select * from Product where
              Description='Toaster'";
            Statement  stat = conn.createStatement();
            ResultSet rs = stat.executeQuery(query);
```

```
            rs.next();
            double price_Toaster=rs.getDouble("Price");
            price_Toaster=price_Toaster-10;
            if(price_Toaster>=0) {
                rs=stat.executeQuery("select * from Product where
                  Description='Hair dryer'");
                rs.next();
                double price_dryer=rs.getDouble("Price");
                price_dryer=price_dryer+10;
                stat.executeUpdate("UPDATE Product SET Price
                  ="+price_Toaster+" where Description='Toaster'");
                stat.executeUpdate("UPDATE Product SET Price
                  ="+price_dryer+" where Description='Hair dryer'");
                conn.commit(); //开始事务处理 Start transaction
              }
            }
            catch(SQLException e){
              try{
                conn.rollback();
                }
              catch(SQLException ex){
                ex.printStackTrace();
              }
          }
          finally   {
            conn.close();
          }
      }
  }
```

12.5 Experimental result（实验结果）

实验结果如图 12-3～图 12-6 所示。

1.

```
Please enter the price : 20
Product_Code=116-064  Description=Toaster  Price=24.95
Product_Code=257-535  Description=Hair dryer  Price=29.95
Please enter the Toaster's price you want to change  : 40
```

Fig12-3. Search data using the price 使用价格搜索数据

	Product Code	Description	Price
▶	116-064	Toaster	40.00
	257-535	Hair dryer	29.95
	643-119	Car vacuum	19.95

Fig12-4. Result of Changed Price 更改价格后的结果

2.

	Product Code	Description	Price
▶	116-064	Toaster	40.00
	257-535	Hair dryer	29.95
	643-119	Car vacuum	19.95

Fig12-5. The initial data 原始数据

	Product Code	Description	Price
▶	116-064	Toaster	30.00
	257-535	Hair dryer	39.95
	643-119	Car vacuum	19.95

Fig12-6. The changed data 改变后数据

第 13 章　Internet Networking（Internet 网络）

13.1　Key points of this chapter（本章要点）

1．The Internet Protocol（Internet 协议）

Data transmission consists of sending/receiving streams of zeroes and ones along the network connection. These zeroes and ones represent two kinds of information: application data, the data that one computer actually wants to send to another, and network protocol data, the data that describe how to reach the intended recipient and how to check for errors and data loss in the transmission. The protocol data follow certain rules set forth by the Internet Protocol Suite, also called TCP/IP, after the two most important protocols in the suite.

数据传输包括发送/接收 0 和 1 在网络流中传输。这些 0 和 1 代表两种信息：应用数据，这类数据是一台计算机与另一台计算机交互的数据；网络协议数据，描述如何到达预期的数据，以及如何检查在传输中的错误和丢失的数据。协议数据遵守互联网协议规定的各项规定，也被称为 TCP/IP，它们是套件中最重要的两个协议。

For the data to arrive at its destination, it must be marked with a destination address. In IP, addresses are denoted by sequences of four numbers, each one byte (that is, between 0 and 255); for example, 202.101.244.101.

数据到达它的目的地，必须用一个目的地地址来标记出来。在 IP 中，地址由四个数值序列来表示，每一个数值是一个字节（位于 0 和 255 之间）；例如，202.101.244.101。

On the Internet, computers can have so called domain names that are easier to remember, such as www.baidu.com. A special service called the Domain Name System (DNS) translates between domain names and Internet addresses.

在互联网上，计算机可以有容易记住的域名，如 www. baidu.com。一种被称为域名系统（DNS）的特殊服务就是实现域名和网络地址之间的转换。

One interesting aspect of IP is that it breaks large chunks of data up into more manageable packets. Each packet is delivered separately, and different packets that are part of the same transmission can take different routes through the Internet. Packets are numbered, and the recipient reassembles them in the correct order.

IP 的一个有趣的地方是它将大块的数据拆分成更多的可管理的数据包。每个数据包单独发送，不同的数据包为同一传输的一部分，可以采取不同的路由通过互联网传送。根据分组编号，在接收方重组它们的正确顺序。

IP has no provision for retrying an incomplete transmission. That is the job of a higher-level protocol, the Transmission Control Protocol (TCP). This protocol attempts reliable delivery of data, with retries if there are failures, and it notifies the sender whether or not the attempt succeeded.

IP 没有提供一个不完整传输重传的机制。这是一个高层协议传输控制协议（TCP）才能实现的功能。该协议提供可靠的数据传递与重传，如果有故障，它会通知发送方是否尝试传送成功。

A computer that is connected to the Internet may have programs for many different purposes. When data are sent to that computer, they need to be marked so that they can be forwarded to the appropriate program. TCP uses port numbers for this purpose. A port number is an integer between 0 and 65,535. For example, by convention, web servers use port 80, whereas mail servers running the Post Office Protocol (POP) use port 110.

连接到互联网的计算机可能有许多不同用途的程序。当数据被发送到该计算机，它们需要被标记，以便可以被转发到适当的程序。TCP 使用端口号来实现这个目的。端口号是介于 0 和 65535 之间的整数。例如，通过约定，Web 服务器使用端口 80，而运行邮件服务（POP）的邮件服务器使用端口 110。

2．Application Level Protocols

1） HTTP: the Hypertext Transfer Protocol, which is used for the World Wide Web. Suppose you type a web address, called a Uniform Resource Locator (URL), such as http:// www.baidu.com /index.html, into the address window of your browser and ask the browser to load the page.

HTTP：超文本传输协议用于万维网。假如你输入一个网址，称为统一资源定位器（URL），如 http://www.baidu.com/index.html，在你的浏览器的地址窗口要求浏览器加载该页面。

HTTP 命令如表 13-1 所示。

Table 13-1　HTTP Commands（HTTP 命令）

Command（命令）	Meaning（含义）
GET	Return the requested item

续表

Command（命令）	Meaning（含义）
HEAD	Request only the header information of an item
OPTIONS	Request communications options of an item
POST	Supply input to a server-side command and return the result
PUT	Store an item on the server
DELETE	Delete an item on the server
TRACE	Trace server communication

2） POP: Post Office Protocol, which is used to download received messages from e-mail servers.

POP：邮局协议用来从电子邮件服务器下载所收到的邮件。

3） SMTP ：To send messages, you use yet another protocol called the Simple Mail Transfer Protocol (SMTP).

要发送消息，可使用另一个称为“简单邮件传输协议”的协议。

3. A Client Program

In the terminology of TCP/IP, there is a socket on each side of the connection (see Figure13-1). In Java, a client establishes a socket with a call.

在 TCP/IP 术语中，在连接的每一侧都有一个 socket（见 figure13-1）。在 Java 中，一个客户端建立一个 socket 发起一次连接。

```
Socket s = new Socket(hostname, portnumber);
```

For example, to connect to the HTTP port of the server baidu.com, you use:

例如，连接到 baidu.com 服务器的 HTTP 端口，使用以下语句：

```
final int HTTP_PORT = 80;
Socket s = new Socket("baidu.com ", HTTP_PORT);
```

Once you have a socket, you obtain its input and output streams:

一旦建立了一个 socket，就可获得输入流和输出流。

```
InputStream instream = s.getInputStream();
OutputStream outstream = s.getOutputStream();
```

When you are done communicating with the server, you should close the socket:

当完成与服务器的通信时，则应该关闭 socket：

```
s.close();
```

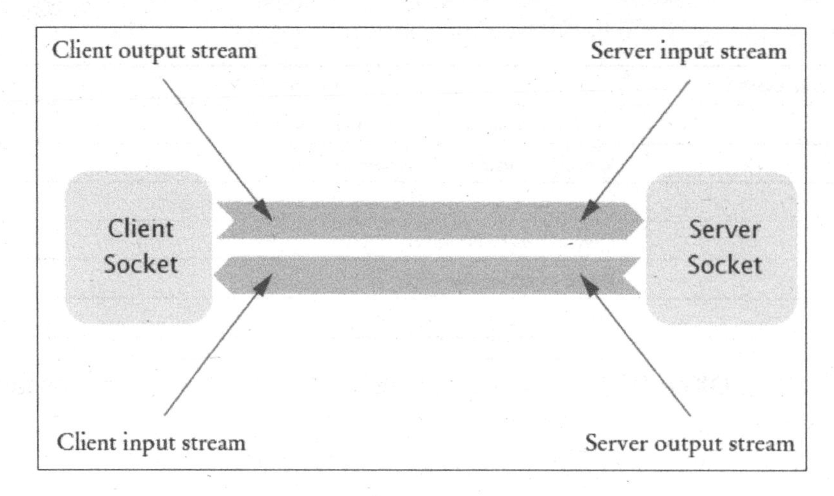

Fig13-1. socket on each side of the connection

4. A Server Program

Whenever you develop a server application, you need to specify some application level protocol that clients can use to interact with the server.

在开发一个服务器应用程序时，需要指定一些应用层的协议供客户端使用来与服务器进行交互。

To listen to incoming connections, you use a server socket. To construct a server socket, you need to supply the port number:

要监听传入的连接，可使用一个服务器套接字。要构建一个服务器套接字，需要提供端口号：

```
ServerSocket server = new ServerSocket(8888);
```

The accept method of the ServerSocket class waits for a client connection. When a Client connects, then the server program obtains a socket through which it communicates with the client:

ServerSocket 类的 accept()方法等待客户端连接。当客户端连接时，服务器程序通过与客户端进行通信，从而获得一个套接字：

```
Socket s = server.accept();
BankService service = new BankService(s, bank);
```

5. URL Connections

The URLConnection class makes it very easy to fetch a file from a web server given the file's URL as a string. First, you construct a URL object from the URL in the familiar format, starting with the http or ftp prefix. Then you use the URL object's openConnection method to get the URLConnection object itself:

如果文件的 URL 为一字符串，URL 连接类则容易从 web 服务器中获取该文件。首先以 http 或 ftp 开始，在熟悉的格式中从 URL 里构建一个 URL 对象。然后使用 URL 对象中的 openConnection 方法来获取 URLConnection 对象：

```
URL u = new URL("http://baidu.com/index.html");
URLConnection connection = u.openConnection();
```

Then you call the getInputStream method to obtain an input stream:

接着调用 getInputStream 方法来获取输入流：

```
InputStream instream = connection.getInputStream();
```

13.2 Example（例题）

1. This program demonstrates how to use a socket to communicate with a web server. Supply the name of the host and the resource on the command-line, for example,

这个程序演示了如何使用一个套接字来与 Web 服务器进行通信。提供了主机的名称和命令行上的资源，例如：

```
import java.io.InputStream;
import java.io.IOException;
import java.io.OutputStream;
import java.io.PrintWriter;
import java.net.Socket;
import java.util.Scanner;
/**
*/
public class WebGet
{
  public static void main(String[] args) throws IOException
  {
    // Get command-line arguments
    String host;
    String resource;
    System.out.println("Getting / from baidu.com");
    host = "baidu.com";
    resource = "/";
    // Open socket
    final int HTTP_PORT = 80;
    Socket s = new Socket(host, HTTP_PORT);
    // Get streams
    InputStream instream = s.getInputStream();
    OutputStream outstream = s.getOutputStream();
    // Turn streams into scanners and writers
    Scanner in = new Scanner(instream);
    PrintWriter out = new PrintWriter(outstream);
    // Send command
```

```
      String command = "GET " + resource + " HTTP/1.1\n"
        + "Host: " + host + "\n\n";
      out.print(command);
      out.flush();
      // Read server response
      while (in.hasNextLine())
      {
        String input = in.nextLine();
        System.out.println(input);
      }
      // Always close the socket at the end
      s.close();
    }
  }
```

2.	Socket object in the client gets the OutputStream object, and then write "Hello From Client" data. Then you can create a server application of the socket, and then call the accept method to listen and access to the client's request socket, you need to print the data which is sended by the client. 从客户端的socket对象中得到OutputStream对象，然后写入“Hello From Client”数据。创建一个服务器端的 socket，然后调用 accept 方法监听并获取客户端的请求 socket，在服务端将输出客户端发送过来的数据。

```
// ClientSocket.java is runed in client(运行在客户端的程序)
import java.io.IOException;
import java.io.OutputStreamWriter;
import java.io.Writer;
import java.net.Socket;
public class ClientSocket {
 public static void main(String args[]) {
   String host = "127.0.0.1";
   int port = 8919;
   try {
    Socket client = new Socket(host, port);
    Writer writer = new OutputStreamWriter(client.getOutputStream());
    writer.write("Hello From Client");
    writer.flush();
    writer.close();
    client.close();
   } catch (IOException e) {
    e.printStackTrace();
   }
  }
}
//ServerClient.java is runed in server.(运行在服务端的程序)
import java.io.InputStreamReader;
```

```
import java.io.Reader;
import java.net.ServerSocket;
import java.net.Socket;
public class ServerClient {
 public static void main(String[] args) {
    int port = 8919;
    try {
      ServerSocket server = new ServerSocket(port);
      Socket socket = server.accept();
      Reader reader = new InputStreamReader(socket.getInputStream());
      char chars[] = new char[1024];
      int len;
      StringBuilder builder = new StringBuilder();
      while ((len=reader.read(chars)) != -1) {
        builder.append(new String(chars, 0, len));
      }
      System.out.println("Receive from client message=: " + builder);
      reader.close();
      socket.close();
      server.close();
    } catch (Exception e) {
      e.printStackTrace();
    }
  }
}
```

3.	This program demonstrates how to use a URL connection to communicate with a web server. Supply the URL on the command line, for example: 这个程序演示了如何使用一个 URL 连接来与一个 Web 服务器进行通信。提供命令行上的 URL，例如：

```
      java URLGet http://baidu.com/index.htm
import java.io.InputStream;
import java.io.IOException;
import java.io.OutputStream;
import java.io.PrintWriter;
import java.net.HttpURLConnection;
import java.net.URL;
import java.net.URLConnection;
import java.util.Scanner;
public class URLGet
{
   public static void main(String[] args) throws IOException
   {
      // Get command line arguments
      String urlString;
```

```
        if (args.length == 1)
        {
          urlString = args[0];
        }
        else
        {
          urlString = "http://baidu.com/";
          System.out.println("Using " + urlString);
        }
        // Open connection
        URL u = new URL(urlString);
        URLConnection connection = u.openConnection();
        // Check if response code is HTTP_OK (200)
        HttpURLConnection httpConnection
             = (HttpURLConnection) connection;
        int code = httpConnection.getResponseCode();
        String message = httpConnection.getResponseMessage();
        System.out.println(code + " " + message);
        if (code != HttpURLConnection.HTTP_OK)
        {
          return;
        }
        // Read server response
        InputStream instream = connection.getInputStream();
        Scanner in = new Scanner(instream);
        while (in.hasNextLine())
        {
          String input = in.nextLine();
          System.out.println(input);
        }
    }
}
```

URL 连接结果如图 13-2 所示。

```
Using http://baidu.com/
200 OK
<html>
<meta http-equiv="refresh" content="0;url=http://www.baidu.com/">
</html>
```

Fig13-2. The result of URLConnection

13.3 Experimental contents（实验内容）

Create a server application that uses the following protocol:

创建使用以下协议的服务器应用程序：

使用的协议如表 13-2 所示。

Table 13-2　uses the following protocol（表 13-2 使用以下协议）

Client Request	Server Response	Meaning
HELLO	greeting	Sends a greeting
ECHO n	n	Echoes n back to the client
COUNT	The number of ECHO requests	Returns a count of the number of ECHO requests by client
QUIT	goodbye	Prints a goodbye and quits the connection

You may choose the greeting and goodbye messages to send. In this exercise you will create three classes:

可以选择问候语和再见消息发送。在这个练习中，创建三个类：

```
SimpleProtocolService
SimpleProtocolServer
SimpleProtocolClient
```

Part of the code of the SimpleProtocolService class has been provided for you:

部分 Simpleprotocolservice 类的代码已经提供给你：

1）

```
public class SimpleProtocolService implements Runnable
{
  public SimpleProtocolService(Socket aSocket) { . . . }
  public void run()
  {
    try
    {
      try
      {
        in = new Scanner(s.getInputStream());
        out = new PrintWriter(s.getOutputStream());
        doService();
      }
      finally
      {
        s.close();
      }
    }
    catch (IOException exception)
    {
      exception.printStackTrace();
    }
  }
  public void doService() throws IOException { . . . }
  public void executeCommand(String command) { . . . }
  . . .
}
```

2）	The server should accept multiple simultaneous connections with different clients. To support multiple simultaneous classes, when the server socket accepts a connection, it should construct a new thread to communicate with that client. What is the code of your SimpleProtocolServer class? 服务器应该可以同时接受多个连接不同客户端的连接。要支持多个同时连接类，当服务器套接字接受一个连接时，它应该构建一个新的线程来与客户端通信。该如何实现 Simpleprotocolserver 类的代码？

3）	Write a client that connects to the server, sends several commands and disconnects. What is the code of your SimpleProtocolClient class? 写一个客户端连接到服务器，发送一些命令然后断开连接。该如何实现 SimpleProtocolClient 类的代码？

13.4 Experimental steps（实验步骤）

1.

```
import java.io.InputStream;
import java.io.IOException;
import java.io.OutputStream;
import java.io.PrintWriter;
import java.net.Socket;
import java.util.Scanner;
public class SimpleProtocolService implements Runnable
{
  private int count;
  private Socket s;
  private Scanner in;
  private PrintWriter out;
  public SimpleProtocolService(Socket aSocket)
  {
    s = aSocket;
    count = 0;
  }
  public void run()
  {
    try
    {
      try
      {
        in = new Scanner(s.getInputStream());
        out = new PrintWriter(s.getOutputStream());
        doService();
      }
```

```
            finally
            {
               s.close();
            }
         }
         catch (IOException exception)
         {
            exception.printStackTrace();
         }
      }
      public void doService() throws IOException
      {
         while (true)
         {
            if (!in.hasNext()) return;
            String command = in.next();
            if (command.equals("QUIT"))
            {
               out.println("Goodbye");
               out.flush();
               return;
            }
            else executeCommand(command);
         }
      }
      public void executeCommand(String command)
      {
         if (command.equals("HELLO"))
         {
            out.println("Hello stranger");
         }
         else if (command.equals("ECHO"))
         {
            int n = in.nextInt();
            count++;
            out.println("" + n);
         }
         else if (command.equals("COUNT"))
         {
            out.println("" + count);
         }
         out.flush();
      }
   }
```

2.

```
   import java.io.IOException;
   import java.net.ServerSocket;
```

```
import java.net.Socket;
public class SimpleProtocolServer
{
   public static void main(String[] args) throws IOException
   {
      final int SBAP_PORT = 8888;
      ServerSocket server = new ServerSocket(SBAP_PORT);
      System.out.println("Waiting for clients to connect...");
      while (true)
      {
         Socket s = server.accept();
         System.out.println("Client connected.");
         SimpleProtocolService service = new SimpleProtocolService(s);
         Thread t = new Thread(service);
         t.start();
      }
   }
}
```

3.

```
import java.io.InputStream;
import java.io.IOException;
import java.io.OutputStream;
import java.io.PrintWriter;
import java.net.Socket;
import java.util.Scanner;
public class SimpleProtocolClient
{
   public static void main(String[] args) throws IOException
   {
      final int SBAP_PORT = 8888;
      Socket s = new Socket("localhost", SBAP_PORT);
      InputStream instream = s.getInputStream();
      OutputStream outstream = s.getOutputStream();
      Scanner in = new Scanner(instream);
      PrintWriter out = new PrintWriter(outstream);
      String command = "HELLO\n";
      System.out.print("Sending: " + command);
      out.print(command);
      out.flush();
      String response = in.nextLine();
      System.out.println("Receiving: " + response);
      command = "ECHO 1\n";
      System.out.print("Sending: " + command);
      out.print(command);
      out.flush();
      response = in.nextLine();
      System.out.println("Receiving: " + response);
```

```
            command = "ECHO 2\n";
            System.out.print("Sending: " + command);
            out.print(command);
            out.flush();
            response = in.nextLine();
            System.out.println("Receiving: " + response);
            command = "COUNT\n";
            System.out.print("Sending: " + command);
            out.print(command);
            out.flush();
            response = in.nextLine();
            System.out.println("Receiving: " + response);
            command = "QUIT\n";
            System.out.print("Sending: " + command);
            out.print(command);
            out.flush();
            s.close();
        }
    }
```

13.5 Experimental result（实验结果）

实验结果如图 13-3 和图 13-4 所示。

```
Sending: HELLO
Receiving: Hello stranger
Sending: ECHO 1
Receiving: 1
Sending: ECHO 2
Receiving: 2
Sending: COUNT
Receiving: 2
Sending: QUIT
```

Fig13-3　the result of Client

```
Waiting for clients to connect...
Client connected.
```

Fig13-4　the result of Server

附录

Java 程序设计试卷 1

1.（15 points）

Prime numbers. Write a program that prompts the user for an integer and then prints out all prime numbers up to that integer. For example, when the user enters 10, the program should print

2

3

5

7

Recall that a number is a prime number if it is not divisible by any number except 1 and itself.

2.（15 points）

A supermarket wants to reward its best customer of each day, showing the customer’ s name on a screen in the supermarket. For that purpose, the store keeps an ArrayList<Customer>. In the Store class, implement methods

```
public void addSale(String customerName, double amount)
//record the sale
public String nameOfBestCustomer()
//return the name of the customer with the largest sale
```

Write a program that prompts the cashier to enter all prices and names, adds them to a Store object, and displays the best customer's name.

3.（15 points）

Design a Corporation class that includes two inner classes, respectively called DevelopingDepartment and SalesDepartment. Then define a Department interface that has a single self-defined method:

```
String work()
```

You should make each department class implement the Department interface.

Similarly you make the Corporation class to implement the Business and Standard interfaces and define several arbitrary（任意的）instance variables in Standard interface. Business interface has a single self-defined method:

```
String makeMoney();
```

It is allowed to be an output statement in the self-defined method.

4.（15 points）

Write an application with a Color menu and menu item labeled “Red” that change the background color of a panel in the center of the frame to red.

Remind(提示):

void actionPerformed(ActionEvent e)

Class ActionEvent	
Modifier and Type	Method and Description
String	getActionCommand():Returns the command string associated with this action.

5.（20 points）

Using map to achieve a shopping cart, Map<Key, Value>, there are two kinds of set implementations, the Java library has two implementations for the Map interface: HashMap and TreeMap. you can put the product as key, the number of goods as value, write a program to realize the function like this.

Input:

Computer 10

Banana 5

Computer 6

Output:

Computer 16

Banana 5

Remind:

Interface Map<K,V>	
Modifier and Type	Method and Description
boolean	containsKey(Object key):Returns true if this map contains a mapping for the specified key.
boolean	containsValue(Object value):Returns true if this map maps one or more keys to the specified value.
V	get(Object key):Returns the value to which the specified key is mapped, or null if this map contains no mapping for the key.
default V	replace(K key, V value):Replaces the entry for the specified key only if it is currently mapped to some value.

6.（20 points）

Write a program WordCount that counts the words in one or more files. Start a new thread for each file. For example, if you call

java WordCount report.txt address.txt Homework.java

then the program might print

address.txt: 1052

Homework.java: 445

report.txt: 2099

Interface Runnable	
Modifier and Type	Method and Description
void	run(): When an object implementing interface Runnable is used to create a thread, starting the thread causes the object's run method to be called in that separately executing thread.

Java 程序设计试卷 1 参考答案

1.（15 points）

```
import java.util.Scanner;
public class PrimeNumbers
{
   public static void main(String[] args)
   {
      Scanner in = new Scanner(System.in);
      System.out.println("Enter n: ");
      int n = in.nextInt();
      for (int currentNumber = 2; currentNumber <= n; currentNumber++)
      {
         boolean primeSoFar = true;
         int testNumber = 2;
         while (primeSoFar && testNumber < currentNumber)
         {
            if (currentNumber % testNumber == 0)
            {
               primeSoFar = false;
            }
            testNumber++;
         }
         if (primeSoFar)
         {
            System.out.println(currentNumber);
         }
      }
   }
}
```

2.（15 points）

```
import java.util.Scanner;
public class BestCustomer {
    public static void main(String argus[]){
        Store s=new Store();
        String customerName;
        Double saleAmount;
        System.out.println("输入 0 则退出");
        do {
```

```
            Scanner in = new Scanner(System.in);
            System.out.println("请输入顾客姓名：");
            customerName=in.nextLine();
            if (customerName.equals("0")) break;
            System.out.println("请输入顾客消费金额：");
            saleAmount=in.nextDouble();
            if (saleAmount==0) break;
            in.nextLine();
            s.addSale(customerName, saleAmount);
        }while (true);
        System.out.println("Best Customer is "+s.nameOfBestCustomer());
    }
}
import java.util.ArrayList;
public class Store {
    private ArrayList<Customer> customerArray = new ArrayList<>();
    public void addSale(String customerName, double amount){
        customerArray.add(new Customer(customerName,amount));
    }
    public String nameOfBestCustomer(){
String name=null;
        Double maxAmount=-1.0;
        for(Customer c:customerArray){
            if(c.getAmount()>maxAmount){
                maxAmount=c.getAmount();
                name=c.getCustomerName();
            }
        }
        return name;
    }
}
class Customer{
    private String customerName;
    private Double amount;
    public Customer(String customerName, Double amount) {
        this.customerName = customerName;
        this.amount = amount;
    }
public Double getAmount() {        return amount;    }
public String getCustomerName() {    return customerName; }
}
```

3.（15 points）

```
public class Corporation implements Bussiness, Standard {
    final String corporationName = "DHLG";
    private DevelopingDepartment devDpt;
    private SalesDepartment saleDpt;
    public Corporation() {
```

```
        devDpt = new DevelopingDepartment();
        saleDpt = new SalesDepartment();
    }
    class DevelopingDepartment implements Department {
        String name="开发部";
        public String work() {   return "开发工作...";    }
    }
    class SalesDepartment implements Department {
        String name="销售部";
        public String work() {   return "销售工作...";    }
    }
    public String makeMoney(){   return "Making a lot of money.";  }
    public void standard() {
        System.out.print(workStandard+"  ");
        System.out.print(dressStandard);
        System.out.println();
    }
    public static void main(String[] args) {
        Corporation cp = new Corporation();
        System.out.println("公司名称:  "+cp.corporationName);
        System.out.print("公司标准:  ");
        cp.standard();
        System.out.println("公司业绩:  "+" "+cp.makeMoney());
        Corporation.DevelopingDepartment cd=cp.new
          DevelopingDepartment();
        Corporation.SalesDepartment cs=cp.new SalesDepartment();
        System.out.println("部门名称:  "+cd.name);
        System.out.println("部门工作:  "+cd.work());
        System.out.println("部门名称:  "+cs.name);
        System.out.println("部门工作:  "+cs.work());
    }
}
interface Department {   String work();    }
interface Bussiness {    String makeMoney();    }
interface Standard {
    String workStandard = "try hard ...";
    String dressStandard = "Beautiful|Handsome";
    void standard();
}
```

4.（15 points）

```
import java.awt.Color;
import java.awt.event.ActionEvent;
import java.awt.event.ActionListener;
import javax.swing.*;
public class MenuOfColor extends JFrame {
    private JPanel panel;
    public MenuOfColor(){
```

```
        JMenu colorMenu=new JMenu("color");
        ActionListener listener=new ClickListener();
        JMenuItem red=new JMenuItem("Red");
        red.addActionListener(listener);
        colorMenu.add(red);
        JMenuBar bar=new JMenuBar();
        bar.add(colorMenu);
        setJMenuBar(bar);
        panel=new JPanel();
        add(panel);
    }
    class ClickListener implements ActionListener{
        @Override
        public void actionPerformed(ActionEvent arg0) {
            String str=arg0.getActionCommand();
            if(str.equals("Red")){
                panel.setBackground(Color.RED);
            }
        }
    }
    public static void main(String[] args) {
        MenuOfColor m1=new MenuOfColor();
        m1.setTitle("color menu");
        m1.setSize(300,200);
        m1.setDefaultCloseOperation(JFrame.EXIT_ON_CLOSE);
        m1.setVisible(true);
    }
}
```

5.（20 points）

```
import java.util.HashMap;
import java.util.Scanner;
public class MapExample {
    public static void main(String[] args) {
        HashMap<String,Integer> map=new HashMap<String,Integer>();
        Scanner in=new Scanner(System.in);
        String name;
        int count;
        System.out.println("Input:");
        String []str=new String[2];
        while(in.hasNextLine())
        {
            String line=in.nextLine();
            if(line.equals(""))
                break;
            str=line.split(" ");
            name=str[0];
            count=Integer.parseInt(str[1]);
```

```
            if(map.containsKey(name))
            {
                int tmp=map.get(name);
                map.replace(name, tmp+count);
            }
            else
            {
                map.put(name,count);
            }
        }
        System.out.println("Output:");
        for(String str1:map.keySet())
        {
            System.out.println(str1+"   "+map.get(str1));
        }
    }
}
```

6.（20 points）

```
import java.util.Scanner;
import java.io.File;
import java.io.FileNotFoundException;
public class WorkRunnable implements Runnable {
    String fileName;
    public WorkRunnable(String fileName) {     this.fileName = fileName; }
    public void run() {
        long count=0;
        Scanner in=null;
        try {
            in = new Scanner(new File(fileName));
            in.useDelimiter("[^a-zA-Z]+");
            while(in.hasNext()){
                in.next();
                count++;
            }
        } catch (FileNotFoundException e) { e.printStackTrace(); }
        System.out.println(fileName+":"+count);
    }
}
public class WordCount {
    public static void main(String[] args) {
        for (String s : args) {
            WorkRunnable wr = new WorkRunnable(s);
            Thread t = new Thread(wr);
            t.start();
        }
    }
}
```

Java 程序设计试卷 2

PART-A

CHOOSE THE BEST ANSWER　　5×2 = 10

1. Which of the answers does the same thing as the following:　value += sum++;

   A. Value = value + sum; sum = sum + 1;

   B. Sum = sum + 1; value = value + sum;

   C. Value = value + sum;

   D. Value = value + ++sum;

2. What is the range of data type short in Java?

   A. -128 to 127　　B. -32768 to 32767

   C. -2147483648 to 2147483647　　D. None of the above

3. What is a variable?

   A. Something that wobbles but doesn't fall down.

   B. Text in a program that the compiler ignores.

   C. A place to store information in a program.

   D. An expression use in Java programming.

4. What is the process of fixing errors called?

   A. Defrosting　　B. Debugging

   C. Decomposing　　D. Compiling

5. What is the output of the following code fragment?

```
for ( int j = 10;  j >  5; j-- ) {
    System.out.print( j + " " );
} System.out.println( );
```

   A. 10 11 12 13 14 15　　B. 9 8 7 6 5 4 3 2 1 0

   C. 10 9 8 7 6 5　　D. 10 9 8 7 6

FILL IN THE BLANKS　　10×2 = 20

1. ________ is the default value of byte data type in Java.

2. Java was first developed in__________

3. ____________ is the full form of JVM

4. __________ is the data type for the number 9.6352. (double)

5. The JDBC-ODBC bridge is _______

6. _______________means subclass extend parent class

7. The extension name of a Java source code file is _______

8. When a program does not want to handle exception, the ______class is used.

9. A _______ is used to separate the hierarchy of the class while declaring an Import statement.

10. ________type of inheritance is not supported by java.

True/False 5×2 = 10

Indicate whether the sentence or statement is true or false. ( )

1. The switch statement does not require a break. ( )

2. A constructor must have the same name as the class it is declared within. ( )

3. Class ClassTwo extends ClassOne means ClassOne is a subclass. ( )

4. The garbage collector will run immediately when the system is out of memory. ( )

5. OOP stands for Object Overloading Programming. ( )

PART-B

ANSWER ALL THE QUESTIONS. 15×4 = 60

1. Write down the syntax for defining interface?

2. What is the difference between the >> and >>> operators?

3. What is the output of following code snippet? Explain Why?

4. Difference between Overloading and Overriding?

5. What is AWT?

6. What are the three OOPs principles and define them?

7. How are the variables declared?

8. What is mean by garbage collection?

9. What is difference between importing "java.applet.Applet" and "java.applet.*;"?

10. What is the difference between 'throw' and 'throws'? And it's application?

11. We cannot use "/* … */" to nest comments in Java. Why?

12. What are the data types used in Java?

13. What is thread.sleep() method?

14. What is the use of keyword static final?

15. What will happen if following code block is compiled/run?

```
int[] iArray = new int[10];
iArray.length = 15;
System.out.println(iArray.length);
```

Java 程序设计试卷 2 参考答案

PART-A

CHOOSE THE BEST ANSWER 5×2 = 10

1. A 2. B 3. C 4. D 5. D

FILL IN THE BLANKS 10×2 = 20

1. Zero
2. 1991
3. Java Virtual Machine
4. Double
5. Multithread
6. Inheritance
7. .java
8. Throws
9. Package
10. Multiple

True/False 5×2 = 10

1. False
2. True
3. False
4. True
5. False

PART-B

ANSWER ALL THE QUESTIONS. 15×4 = 60

1.

Syntax for defining Interface:

Interface declaration have public access specifier, Interface name. Interface body has declarations and methods.

Syntax:

```
public interface GroupOfInterfaces extends Interface1, Interface2, ..{
    //declaration
    ....
    //methods
    ....
    }
```

2.

The >> operator carries the sign bit when shifting right. The >>> zero-fills bits that have been shifted out.

3.

Explanation:

It is binary representation of 26.

4.

Method overloading increases the readability of the program.

Method overriding provides the specific implementation of the method that is already provided by its super class parameter must be different in case of overloading, parameter must be same in case of overriding.

5.

Java AWT-Abstract Window Toolkit is an API used to develop GUI or Window Based Application in Java.

AWT contains large number of classes and methods supporting creation of GUI applications.

It forms the foundation of Java Swing.

It is platform independent and heavyweight.

6.

Encapsulation, Inheritance and Polymorphism are the three OOPs Principles.

Encapsulation: It is the Mechanism that binds together code and the data it manipulates, and keeps both safe from outside interference and misuse.

Inheritance: It is the process by which one object acquires the properties of another object.

Polymorphism: It is a feature that allows one interface to be used for a general class of actions.

7.

Variables can be declared anywhere in the method definition and can be initialized during their declaration. They are commonly declared before usage at the beginning of the definition. Variables with the same data type can be declared together. Local variables must be given a value before usage.

8.

When an object is no longer referred to by any variable, Java automatically reclaims memory used by that object. This is known as garbage collection.

9.

"java.applet.Applet" will import only the class Applet from the package java.applet. Whereas "java.applet.*" will import all the classes from java.applet package.

10.

Exceptions that are thrown by java runtime systems can be handled by Try and catch blocks.With throw exception we can handle the exceptions thrown by the program itself. If a method is capable of causing an exception that it does not handle, it must specify this behavior so the callers of the method can guard against that exception.

11.

We cannot simply deactivate a code simply by using “/*　　*/ ”. Because the deactivated code may also contain the delimiter, “*/”in it.

12.

We have eight primitive data types in java. They are:

1). integer 2).short integer 3). Long integer 4). Byte 5).float

6). Double 7).character type 8). Boolean type

13.

The thread.sleep() method is a static method Thread class.

This method temporarily stops the activity of the currently running method.

In this method, we can also specify the time (in milliseconds) to keep the thread sleep

14.

The keyword *static final* is used to denote a constant. It indicates that you can assign to the variable once, then its value is set once and for all.

Eg: public class Constant

```
{
 public static void main(String args[]) {
   double width=8.3;
   double height=2.5;
  System.out.println("PaperSize="+width*cm+height*cm);
   }
   public static final double cm=2.54;
}
```

15.

Compilation Error

Explanation: Once an array is created, it is not possible to change the length of the array.

参考文献

[1] 何月顺．双语版 Java 程序设计．北京：电子工业出版社，2012．

[2] Raoul-Gabriel Urma, Mario Fusco, and Alan Mycroft, Java 8 in Action, Manning Publications Co., 2015．

[3] Paul Deitel, Harvey Deitel, Java How To Program(10th Edition), Pearson, 2015．

[4] C. Horstmann. Big Java Late Objects ,Wiley, 2014．

[5] Y. Daniel Liang, Introduction to Java Programming, Comprehensive Version (10th Edition), Pearson, 2015．

[6] 陈国君．Java 程序设计基础（第 5 版）．北京：清华大学出版社，2015．

[7] 黄文海．Java 多线程编程实战指南（设计模式篇）．北京：电子工业出版社，2015．

[8] 葛一鸣．实战 Java 高并发程序设计．北京：电子工业出版社，2015．

[9] 邹林达．Java 程序设计基础（第 4 版）实验指导．北京：清华大学出版社，2014．

[10] 刘小晶．数据结构实例解析与实验指导——Java 语言描述．北京：清华大学出版社，2013．

[11] 范玫．Java 语言面向对象程序设计（第 2 版）实验指导及习题解答．北京：清华大学出版社，2015．

反侵权盗版声明